Michael Pollan

Second Nature

"Deft and often dazzling . . . about gardening, but only in the same way that Dante's *Divine Comedy* is about getting lost in the woods. . . . I know no book on gardening that is quite as illuminating and fascinating as this one."

—Allen Larcy, *The New York Times*

"You don't have to be a gardener to love *Second Nature*. Pollan is a marvelous essayist: indeed, he is to gardens what Lewis Thomas is to medicine—expert, witty, and original."

—Frances FitzGerald

"*Second Nature* is to gardening what Izaak Walton's *The Compleat Angler* is to fishing. Combining humor, natural description, and advice, it's not so much about compost, seeds, seasons, and pests as it is about human nature."

—Thomas D'Evelyn, *The Christian Science Monitor*

"Wonderful . . . Pollan brings the shrewd eye of a social historian. . . . Most things in Pollan's book work upward toward metaphor—even though he takes care to root every metaphor in aerated soil, rich with the compost of organic experience."

—Richard Dyer, *The Boston Globe*

"He's written a book about gardening that even non-gardeners might want to read. . . . Pollan can still remember that there are readers of intelligence and curiosity whose gardening habits amount to no more than a stroll through the yard every month or so to see what's died."

—Malcolm Jones Jr., *Newsweek*

"As a non-gardener, I never expected to stay up late and laugh out loud at a book like this, but I've been permanently Pollan-ated."

—Christopher Buckley, *Vanity Fair*

"Pollan is a hybrid—a gardener-philosopher-humorist-polemicist who has written a book that manages to amuse while it muses, a book that lures even the non-gardener into the physical and metaphysical garden."

—Jocelyn McClurg, *The Harford Courant*

"As a gardener, I read this charming and ultimately profound book with admiration and, I must admit, some envy; as a writer, with pleasure."
—Witold Rybczynski

"Articulate and funny, *Second Nature* is my idea of a gardening book. . . . What [Pollan] sets out as his aim, and finds . . . is something outside the usual American alternatives, which seem to consist of either raping the land or sealing it away in preserve where no other can touch it. That place is a garden."
—*Chicago Tribune*

"Wonderful writing . . . These elegant, lively, and impeccably crafted essays [offer] us a provocative new way of approaching our environmental problems. . . . At a time when it seems we must choose between unchecked development or no development at all, Pollan's idea of the world as a garden could offer us a way out of the wilderness."
—Inga Saffron, *The Philadelphia Inquirer*

"The best gardening book I have read in memory, perhaps ever . . . [Pollan's] essays are funny and profound, elegant and basic. . . . [*Second Nature*] is the story of Pollan's effort to coexist with nature, forging a middle ground between allowing nature to fulfill its tendency to run rampant . . . and restraining it completely a la American suburbia and its broad, picture-perfect lawns."
—Nancy Brachey, *The Charlotte Observer*

"A bounty of food for thought . . . [Pollan] takes a deep look at our spiritual, ethical, environmental connection to the garden and land itself, philosophically exploring our attitude toward nature and wilderness and how each should be tended."
—Karen A. Cleath, *The Tampa Tribune*

"A serious undertaking and an important book, a reasoned argument with Thoreau and others about the wilderness ethic and how much or little it can tell us about what our attitude toward nature ought to be."
—Christopher Reed, *Horticulture*

"A book for the nightstand, not the coffee table or potting shed . . . A merry read with a major message: garden nirvana may be only a state of mind."
—*Elle Décor*

"I *love this book* . . . the author is witty and spirited. Everything he writes reveals as much of himself as it does of gardening."
—*Hortus*

"Unlike any other gardening book that will appear this season or for many seasons to come. While it may, or may not, help you grow things more successfully, it will almost surely make you think better, even if you have never been tempted to bring ordered growth out of the earth. It is a book for readers first and gardeners second; a book that is wise and funny and an effortless pleasure to read."

—*American Way*

"An important and profoundly original book . . . A well-developed philosophy of life and nature in a technological world."

—*Kirkus Reviews*

"Quirky and pleasing . . . The debut of a fresh and provocative voice in American writing."

—Annie Dillard

Also by Michael Pollan:

A Place of My Own: The Education of an Amateur Builder

The Botany of Desire: A Plant's-Eye View of the World

Second Nature

Second Nature

A GARDENER'S EDUCATION

Michael Pollan

GROVE PRESS
New York

The author wishes to thank the editors of *Harper's Magazine* and *The New York Times Magazine*, where versions of several chapters first appeared.

Printed in the United States of America
Published simultaneously in Canada

Library of Congress Cataloging-in-Publication Data
Pollan, Michael.
Second nature / by Michael Pollan.
ISBN-10: 0-8021-4011-4
ISBN-13: 978-0-8021-4011-1
1. Gardening. 2. Gardening—Philosophy. I. Title.
SB455.P58 1991 635.9--dc20 90-20960

Grove Press
an imprint of Grove/Atlantic, Inc.
841 Broadway
New York, NY 10003

Distributed by Publishers Group West

www.groveatlantic.com

08 09 10 11 12 12 11 10 9 8 7

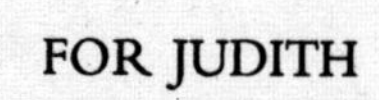

FOR JUDITH

Contents

Introduction

This book is the story of my education in the garden. The garden in question is actually two, one more or less imaginary, the other insistently real. The first is the garden of books and memories, that dreamed-of outdoor utopia, gnat-free and ever in bloom, where nature answers to our wishes and we imagine feeling perfectly at home. The second garden is an actual place, consisting of the five acres of rocky, intractable hillside in the town of Cornwall, Connecticut, that I have been struggling to cultivate for the past seven years. Much separates these two gardens, though every year I bring them a little more closely into alignment.

Both of these gardens have had a lot to teach me, and not only, as it turned out, about gardening. For I soon came to the realization that I would not learn to garden very well before I'd also learned about a few other things: about my proper place in nature (was I within my rights to murder the woodchuck that had been sacking my vegetable garden all spring?); about the somewhat peculiar attitudes toward the land that an American is born with (why is it the neighbors have taken such a keen interest in the state of my lawn?); about the troubled borders between nature and culture; and about the experience of place, the moral implications of landscape design, and several other questions that the wish to harvest a few decent tomatoes had not prepared me for. It may be my nature to complicate matters unduly, to search for large meanings in small things, but it did seem that there was a lot more going on in the garden than I'd expected to find.

I began gardening for the same reasons people usually do: for the satisfaction of pulling bunches of carrots from one's own ground; the desire to make a patch of land more hospitable or productive; the urge to recover a place remembered from childhood, and the basic need to keep the forest from swallowing up one's house. When my wife and I bought our first place in 1983—a sliver of a derelict dairy farm on the eastern edge of the Housatonic Valley—we had been living in Manhattan for some time, in an apartment that receives approximately ninety minutes of sunlight a day, and the prospect of growing a few flowers and vegetables seemed exotic. There was also the matter of the advancing forest, which did in fact threaten to engulf our house, a little cape that had been assembled from a Sears, Roebuck kit in 1929. I had to do *something*—either mow the weed patch that passed for a lawn, or put in a real garden—if I hoped to keep the woods at bay.

So I guess you could say the forest made me do it. But there was also, mixed in with my motives, the recollected satisfactions of childhood gardens. Growing up on Long Island in the early 1960s, I'd cared for a succession of pocket gardens in various corners of my parents' suburban plot, and had spent many Saturdays helping out my grandfather in the much grander garden he tended a few miles away (Chapter 1 is a reminiscence of these places). Now I had some ground of my own, and gardening it seemed a natural way to spend my weekends, something I might even have a knack for.

Judith had other ideas. Though her position would eventually soften, she started out a sworn enemy of gardening, having been forced as a child to do yard work. I think she was also less troubled by the derelict parts of our property than I was, finding beauty in the march of brush across an abandoned hay field, or the rank, top-heavy growth of an apple tree in need of hard pruning. So she began making landscape paintings and I, with somewhat less striking results, began making landscapes.

It wasn't very long before I discovered I was ill-prepared for the work I'd taken on. The local New England landscape—a patchwork of abandoned farms swiftly being overtaken by second-growth forest—proved far less amenable to my plans for it than the tame suburban plots of my childhood had. Here were large and rapacious animals, hegemonies of weeds, a few billion examples of every insect in the field guide, killing frosts in June and September, and boulders of inconceivable weight and number. But there were obstacles of a very different kind that proved just as vexing: the unexamined attitudes toward nature that I'd brought with me to the garden.

Like most Americans out-of-doors, I was a child of Thoreau. But the ways of seeing nature I'd inherited from him, and the whole tradition of nature writing he inspired, seemed not to fit my experiences. In confronting the local woodchucks, or deciding whether I was obliged to mow my lawn, or how liberal I could afford to be with respect to weeds, I was deep in nature, surely, but my feelings about it, although strong, were something other than romantic, or worshipful. When one summer I came across Emerson's argument that "weeds" (just then strangling my annuals) were nothing more than a defect of my perception, I felt a certain cognitive dissonance. Everybody wrote about how to *be* in nature, what sorts of perceptions to have, but nobody about how to *act* there. Yet the gardener, unlike the naturalist, has to, indeed *wants* to, act.

Now it is true that there are countless volumes of practical advice available to the perplexed gardener, but I felt the need for some philosophical guidance as well. Before I firebomb a woodchuck burrow, I like to have a bit of theory under my belt. Yet for the most part, Americans who write about nature don't write about the garden—about man-made landscapes and the processes of their making. This is an odd omission, for although gardening may not at first seem to hold the drama or grandeur of, say, climbing mountains, it is gardening that gives most of us our most direct and intimate experience of nature—of its satisfactions, fragility, and power.

Yet traditionally, when we have wanted to think about our relationship to nature, we have gone to the wilderness, to places untouched by man. Thoreau, in fact, was the last important American writer on nature to have anything to say about gardening. He planted a bean field at Walden and devoted a chapter to his experiences in it. But the bean field (which I talk about in my chapter on weeds) got Thoreau into all sorts of trouble. His romance of wild nature left him feeling guilty about discriminating against weeds (he rails against the need for such "invidious distinctions") and he couldn't see why he was any more entitled to the harvest of his garden than the resident woodchucks and birds. Badly tangled up in contradictions between his needs and nature's prerogatives, Thoreau had to forsake the bean field, eventually declaring that he would prefer the most dismal swamp to any garden. With that declaration, the garden was essentially banished from American writing on nature.

I think this is unfortunate, and not just because I happen to stand in need of sound advice in the garden. Americans have a deeply ingrained habit of seeing nature and culture as irreconcilably opposed; we automatically assume that whenever one gains, the other must lose. Forced to choose, we usually opt for nature (at least in our books). This choice, which I believe is a false one, is what led Thoreau and his descendants out of the garden. To be sure, there is much to be learned in the wilderness; our unsurpassed tradition of nature writing is sufficient proof of that. But my experience in the garden leads me to believe that there are many important things about our relationship to nature that *cannot* be learned in the wild. For one thing, we need, and now more than ever, to learn how to use nature without damaging it. That probably can't be done as long as we continue to think of nature and culture simply as antagonists. So how do we begin to find some middle ground between the two? To provide for our needs and desires without diminishing nature? The premise of this book is that the place to look for some of the answers to these questions may not be in the woods, but in the garden.

Though this book is not a polemic, it is full of argument: between me and this vexing piece of land, and also between me and some of the traditional ways of looking at nature in America; I find I spend a lot of time arguing with Thoreau. Many of these arguments don't get settled; this book is an exercise in discovery rather than truth telling. It is, as I say, the story of an education, and, as will be clear from the high incidence of folly in these pages, I remain more pupil than teacher. I know more at the end of my narrative than I did at the beginning, and for the most part I have followed the logic of my experiences, as they unfolded season by season, rather than that of any thesis. Even so, there is, I think, threading through this book (and spelled out in some detail in Chapter 10), a single underlying argument: that the idea of a garden—as a place, both real and metaphorical, where nature and culture can be wedded in a way that can benefit both—may be as useful to us today as the idea of wilderness has been in the past. This might strike readers as a rather unfashionably optimistic notion. In fact I share the general sense of alarm about our environment; I do not, however, share the gathering sense of despair. I find, in the garden, some grounds for hope.

What are my qualifications to write such a book? Certainly I am no expert—not on gardening, or nature, or much of anything else, for that matter. This is very much the enterprise of an amateur. My sole qualification (if it may be called that) is the wager I decided to make at the beginning of this project: that gardening might be worth taking seriously, and that, closely attended to, it might yield some good stories and helpful ideas. Yet I suspect that once I began to garden, this book was probably inevitable. As most gardeners will testify, the desire to make a garden is often followed by a desire to write down your experiences there—in a notebook, or a letter to a friend who gardens, or if, like me, you make your living by words, in a book. Writing and gardening, these two ways of rendering the world in rows, have a great deal in common. In my part of the country, there comes each year one long and occasionally fruitful season when gardening takes place strictly on paper and in the imagination. This book is how I've spent the last few such seasons in my garden.

I have been fortunate to have the help and encouragement of many people in this undertaking, but it has been Judith's that really made this book possible. Her initial reluctance in the garden eventually gave way to a catching enthusiasm, and we have been close partners in all that follows—in the book as well as the garden. Neither would be of any account without her eye and ear and intelligence.

I'm especially grateful for the generosity and insight of Mark Edmundson who, despite a complete lack of interest in anything having to do with gardening, gave me invaluable criticism and advice at every stage. I also had crucial support from Mark Danner and my colleagues at *Harper's Magazine.* My thanks, too, to Amanda Urban, Ann Godoff, and Carl Navarre, for having faith in this project long before there was any good reason to.

There are a few other people whose influence, unbeknownst to them, has been decisive. This book grows out of my experiences in the library as well as in the garden, and I would not have gotten very far had I never encountered the work of Wendell Berry, Frederick Turner, Eleanor Perényi, Richard Rorty, William Cronon, and J. B. Jackson. Different as these writers are, they are all pioneers on the frontier of nature and culture, and that makes them superb, if perhaps unwitting, guides to the garden.

CHAPTER 1

Two Gardens

My first garden was a place no grown-up ever knew about, even though it was in the backyard of a quarter-acre suburban plot. Behind our house in Farmingdale, on Long Island, stood a rough hedge of lilac and forsythia that had been planted to hide the neighbor's slat wood fence. My garden, which I shared with my sister and our friends, consisted of the strip of unplanted ground between the hedge and the fence. I say that no grown-up knew about it because, in an adult's picture of this landscape, the hedge runs flush against the fence. To a four-year-old, though, the space made by the vaulting branches of a forsythia is as grand as the inside of a cathedral, and there is room enough for a world between a lilac and a wall. Whenever I needed to be out of range of adult radar, I'd crawl beneath the forsythia's arches, squeeze between two lilac bushes, and find myself safe and alone in my own green room.

I think of this place today as a garden not only because it offered an enclosed and privileged space out-of-doors but also because it was here that I first actually grew something. Most of the pictures I can retrieve from that time are sketchy and brittle, but this one unspools like a strip of celluloid. It must be September. I am by myself behind the hedge, maybe hiding from my sister or just poking around, when I catch sight of a stippled green football sitting in a tangle of vines and broad leaves. It's a watermelon. The feeling is of finding treasure—a right-angled change of fortune, an unexpected boon. Then I make the big connection between this melon and a seed I planted, or at least spit out and buried, months before: *I made this happen.* For a moment I'm torn between leaving the melon to ripen and the surging desire to publicize my achievement: *Mom has got to see this.* So I break the cord attaching the melon to the vine, cradle it in my arms and run for the house, screaming my head off the whole way. The watermelon weighs a ton, though, and just as I hit the back steps I lose my balance. The melon squirts from my arms and smashes in a pink explosion on the cement.

Watermelon perfume fills the air and then the memory stalls. I can't remember but I must have cried—to see so fine a triumph snatched away, to feel Humpty-Dumpty suddenly crash onto my four-year-old conscience. Memories of one kind or another play around the edges of every garden, giving them much of their resonance and savor. I've spent thousands of hours in the garden since that afternoon, and there is perhaps some sense in which all this time has been spent trying to recover that watermelon and the flush of pride that attended its discovery.

I can't recall whether I tried to salvage any part of the melon to show my father when he got home from work, but I can assume he would not have been greatly impressed. My father was not much for gardening, and the postage-stamp yard of our ranch house showed it. The lawn was patchy and always in need of a mowing, the hedges were unclipped and scraggly, and in summer hordes of Japanese beetles dined on our rosebushes without challenge. My father was a Bronx boy who had been swept to the suburbs in the postwar migration. Buying a house with a yard on Long Island was simply what you did then, part of how you said who you were when you were a lawyer or a dentist (he was a lawyer) just starting out in the fifties. Certainly it was no great love of fresh air that drove him from the city. I have a few memories of my father standing with his Salem and a highball glass on the concrete patio behind the house, but, with a single exception I will come to, not one of him out in the yard mowing the lawn or pulling weeds or otherwise acting the part of a suburban dad.

I remember him as strictly an indoor dad, moving around the house in his year-round uniform of button-down shirt, black socks and tie shoes, and boxer shorts. Maybe it was the fact that he hated to wear pants that kept him indoors, or perhaps the boxers were a way to avoid having to go outside. Either way, my mother was left with the choice of her husband doing the yard work in his underwear, or not doing it at all, which in the suburbs is not much of a choice. So while the boxers kept Dad pinned to the kitchen table, the yard steadily deteriorated to the point where it became something of a neighborhood and family scandal.

My mother's father lived a few miles away in Babylon, in a big house with beautiful, manicured gardens, and the condition of our yard could be counted on to make him crazy—something it may well have been calculated to do. My grandfather was a somewhat overbearing patriarch whom my father could not stand. Grandpa, who would live to be ninety-six, had come to Long Island from Russia shortly before the First World War. Starting out with nothing, selling vegetables from a horse wagon, he eventually built a fortune, first in the produce business and later in real estate. In choosing my father, my mother had married a notch or two beneath her station, and Grandpa made it his business to minimize his eldest daughter's sacrifice—or, looked at from another angle, to highlight my father's shortcomings. This meant giving my father large quantities of unsolicited career advice, unsolicited business opportunities (invariably bum deals, according to my father), and unsolicited landscape services.

In the same way some people send flowers, Grandpa sent whole gardens. These usually arrived unexpectedly, by truck caravan. Two or three flatbeds appeared at the curb and a crew of Italian laborers fanned out across the property to execute whatever new plan Grandpa had dreamed up. One time he sent a rose garden that ran the length of our property, from curb to back fence. But it wasn't enough to send the rosebushes: Grandpa held my father's very soil in low esteem; no plant of *his* could be expected to grow in it. So he had his men dig a fifty-foot trench three feet wide and a foot deep, remove the soil by hand and then replace it with soil trucked in from his own garden. This way the roses (which also came from Grandpa's garden) would suffer no undue stress and my father's poor, neglected soil would be at least partly redeemed. Sometimes it seemed as if my grandfather was bent on replacing every bit of earth around our house, a square foot at a time.

Now any good gardener cares as much about soil as plants, but my grandfather's obsession with this particular patch of earth probably went deeper than that. No doubt my father, who was the first in his family to own his own house, viewed his father-in-law's desire to replace our soil with his own as a challenge to the very ground on which his independence stood. And maybe there was something to this: Grandpa had given my parents the money for the down payment ($4,000; the house had cost $11,000), and, like most of his gifts, this one was not unencumbered. The unsolicited landscaping services, like Grandpa's habit of occasionally pounding on the house's walls as if to check on its upkeep, suggest that his feelings about our house were more than a little proprietary. It was as a landlord that Grandpa felt most comfortable in the world, and as long as my father declined to think of himself as a tenant, they were bound not to get along.

But probably his concern for our soil was also an extension of his genuine and deeply felt love of land. I don't mean love of *the* land, in the nature-lover's sense. *The* land is abstract and in some final sense unpossessable by any individual. Grandpa loved land as a reliable if somewhat mystical source of private wealth. No matter what happened in the world, no matter what folly the government perpetrated, land could be counted on to hold and multiply its value. At the worst a plot could yield a marketable crop and, at least on Long Island for most of this century, it could almost certainly be resold for a profit. "They can print more money," he liked to say, "and they can print new stocks and bonds, but they can't print more land."

In his mind, the Old World peasant and the real estate developer existed side by side; he was both men and perceived no contradiction. Each looked at a piece of land and saw potential wealth: it made no difference that one saw a field of potatoes and the other a housing development. Grandpa could be perfectly happy spending his mornings tenderly cultivating the land and his afternoons despoiling it. Thoreau, planting his bean field, said he aimed to make the earth "speak beans." Some days my grandfather made the earth speak vegetables; other days it was shopping centers.

Grandpa started out in the teens wholesaling produce in Suffolk County, which was mostly farmland then. He would buy fruits and vegetables from the farmers and sell them to restaurants and, later, to the military bases that sprang up on Long Island during the war. He managed to make money straight through the Depression, and used his savings to buy farmland at Depression prices. When after the Second World War the suburbs started to boom, he saw his opportunity. Suffolk County was generally considered too far from the city for commuters, but Grandpa was confident that sooner or later the suburban tide would reach his shore. His faith in the area was so emphatic that (according to his obituary in *Newsday*) he was known in business circles as Mr. Suffolk.

Grandpa worked the leading edge of the suburban advance, speculating in the land that suburbanization was steadily translating from farm into tract house and shopping center. He grasped the powerful impulses that drove New Yorkers farther and farther out east because he shared them. There was the fear and contempt for city ways—the usual gloss on the suburban outlook—but there was also a nobler motive: to build a middle-class utopia, impelled by a Jeffersonian hunger for independence and a drive to create an ideal world for one's children. The suburbs, where you could keep one foot on the land and the other in the city, was without a doubt the best way to live, and Grandpa possessed an almost evangelical faith that we would all live this way eventually. Every time he bought a hundred acres of North Fork potato field, he knew it was only a matter of time before its utopian destiny would be fulfilled. Grandpa had nothing against potatoes, but who could deny that the ultimate Long Island crop was a suburban development? The fact that every home in that development could have a patch of potatoes in the backyard was proof that progress had no cost.

His own suburban utopia was a sprawling ranch house on five acres of waterfront in Babylon, on the south shore. My grandfather had enough money to live nearly anywhere, and for a time the family lived in a very grand mansion in Westbury. But he preferred to live in one of Long Island's new developments, and after his children were grown he and my grandmother moved into one where the fancy homes on their big plots nevertheless hewed to the dictates of middle-class suburban taste. The houses were set well back from the road and their massive expanses of unhedged front lawn ran together to create the impression of a single parklike landscape. Here in front of each house was at least an acre of land on which no one but the hired gardener ever stepped, an extravagance of unused acreage that must have rubbed against Grandpa's grain. But front yards in the suburbs are supposed to contribute to a kind of visual commons, and to honor this convention, Grandpa was willing to deny himself the satisfaction of fully exploiting an entire acre of prime real estate.

At least until I was a teenager, visits to Grandma and Grandpa's were always sweet occasions. The anticipation would start to build as we turned onto Peninsula Drive and began the long, slow ride through that Great Common Lawn, a perfection of green relieved only by evergreen punctuation marks and the fine curves of driveways drawn in jet-black asphalt. Eager as we were to get there, we always made Mom slow down (Dad hardly ever made the trip) in the hope of spotting the one celebrity who lived on my grandparents' road: Bob Keeshan, known to every child of that time as Captain Kangaroo. One time we did see the Captain, dressed in his civilian clothes, digging in his garden.

There is something about a lush, fresh-cut lawn that compels children to break into a sprint, and after the long ride we couldn't wait to spill out of the station wagon and fan out across the backyard. The grass always seemed to have a fresh crew cut, and it was so springy and uniform that you wanted to run your hand across it and bring your face close. My sisters could spend the whole afternoon practicing their cartwheels on it, but sooner or later Grandma would lure them indoors, into what was emphatically her realm. Except for the garage and a small den with a TV, where Grandpa passed rainy days stretched out on the sofa, the house brimmed with grandmotherness: glass cases full of tiny ceramic figurines, billowy pink chiffon curtains, dressing tables with crystal atomizers and silver hairbrushes, lacquered boxes stuffed with earrings, ornately framed portraits of my mother and aunt. I remember it as a very queenly place, a suburban Versailles, and it absorbed my sisters for hours at a time.

Grandpa's realm was outside, where he and his gardener, Andy, had made what I judged a paradise. Beginning at the driveway, the lawn described a broad, curving avenue that wound around the back of the house. On one side of it was the flagstone patio and rock garden, and on the other a wilder area planted with shrubs and small trees; this enclosed the backyard, screening it from the bay. A stepping-stone path conducted you through this area, passing beneath a small rose arbor and issuing with an unfailingly pleasing surprise onto the bright white beach. Plunked in the middle of the lawn was a gazebo, a silly confection of a building that was hardly ever used. Arrayed around it in a neat crescent was a collection of the latest roses: enormous blooms on spindly stems with names like Chrysler and Eisenhower and Peace. In June they looked like members of a small orchestra, performing for visitors in the gazebo.

The area between the lawn and the beach was twenty or thirty feet deep, thickly planted, and it formed a kind of wilderness we could explore out of sight of the adults on the patio. Here were mature rhododendrons and fruit trees, including a famous peach that Grandpa was said to have planted from seed. It was an impressive tree, too, weighed down in late summer with bushels of fruit. The tree was a dwarf, so we could reach the downy yellow globes ourselves. Hoping to repeat Grandpa's achievement, we carefully buried the pit of every peach we ate. (Probably it was his example that inspired my experiment with watermelon seeds.) But ripe fruit was only one of the surprises of Grandpa's wild garden. There was another we always looked for, only sometimes found. Creeping among the rhododendrons and dwarf trees, we would on lucky days come upon a small, shaded glade where, on a low mound, a concrete statue stood. It was a boy with his hand on his penis, peeing. This scandalous little scene never failed to set off peals of laughter when we were in a group; alone, the feelings were more complicated. In one way or another Eros operates in every garden; here was where he held sway in Grandpa's.

Back out in daylight, you could continue along the avenue of lawn until you came to an area of formal hedges clipped as tall as a ten-year-old, and forming an alley perhaps ten feet wide and forty feet long. At one end was a regulation-size shuffleboard court paved in sleek, painted concrete (it felt cool to bare feet all summer), and, at the other, a pair of horseshoe pins. Some visits these games held my interest for a while, but usually I made straight for the break in the hedge that gave onto what was unquestionably my favorite and my grandfather's proudest part of the garden—indeed, the only part of the property I ever heard anybody call a garden: his vegetable garden.

Vegetables had given Grandpa his earliest success, and the older he got, the more devoted to them he became. Eventually care of the ornamental gardens fell to Andy, and Grandpa spent the better part of his days among the vegetables, each spring adding to the garden and subtracting from the lawn. It's quite possible that, had Grandpa lived another twenty years, his suburban spread would have reverted entirely to farm. As it was, Grandpa had at least a half-acre planted in vegetables—virtually a truck farm, and a totally unreasonable garden for an elderly couple. I have a photograph of him from the seventies, standing proudly among his vegetables in his double-knits, and I can count more than twenty-five tomato plants and at least a dozen zucchini plants. You can't see the corn—row upon row of sweet corn—or the string beans, cucumbers, cantaloupes, peppers, and onions, but there had to be enough here to supply a farm stand.

The garden was bordered by a curving brick kneewall that ran right along the water, a location that ensured a long growing season since the bay held heat well into the fall, forestalling frost. Grandpa could afford to be extravagant with space, and no two plants in his garden ever touched one another. I don't think a more meticulous vegetable garden ever existed; my grandfather hoed it every morning, and no weed dared raise its head above that black, loamy floor. Grandpa brought the same precision to the planting of string beans and tomatoes that Le Nôtre brought to the planting of chestnut trees in the Tuileries. The rows, which followed the curves of the garden wall, might as well have been laid out by a surveyor, and the space between each plant was uniform and exact. Taken as a whole, the garden looked like nothing so much as a scale model for one of the latest suburban developments: the rows were roads, and each freestanding vegetable plant was a single-family house. Here in the garden one of the unacknowledged contradictions of Grandpa's life was symbolically resolved: the farmer and the developer became one.

But what could have possessed my grandfather to plant such a *big* vegetable garden? Even cooking and canning and pickling at her furious clip, there was no way my grandmother could keep up with his garden's vast daily yield. Eventually she cracked and went on strike: she refused to process any more of his harvest, and true to her word, never again pickled a cucumber or canned a tomato. But even then he would not be deterred, and the garden continued to expand.

I suspect that this crisis of overproduction suited Grandpa just fine. He was foremost a capitalist and, to borrow a pair of terms from Marx, was ultimately less interested in the use-value of his produce than in its exchange-value. I don't mean to suggest that he took no immediate pleasure in his vegetables; his tomatoes, especially, pleased him enormously. He liked to slice his beefsteaks into thick pink slabs and go at them with a knife and fork. Watching him dine on one, you understood immediately how a tomato could come to be named for a cut of meat. "Sweetasugar," he would announce between bites, his accent mushing the three words together into one incantatory sound. Of course he would say the same thing about his Bermuda onions, his corn, even his bell peppers. Grandpa's vocabulary of English superlatives was limited, and "sweet-asugar" was the highest compliment you could pay a vegetable.

Eating beefsteaks was one pleasure, but calculating their market value and giving them away was even better. Having spent many years in the produce business, Grandpa had set aside a place in his mind where he maintained the current retail price of every vegetable in the supermarket; even in his nineties he would drop by the Waldbaum's produce section from time to time to update his mental price list. Harvesting alongside him, I can remember Grandpa holding a tomato aloft and, instead of exclaiming over its size or perfect color, he'd quote its market price: *Thirty-nine cents a pound!* (Whatever the amount, it was always an outrage.) Probably when he gazed out over his garden he could see in his mind's eye those little white placards stapled to tongue depressors listing the going per-pound price of every crop. And given the speed with which he could tally a column of figures in his head, I am sure he could mentally translate the entire garden into U.S. currency in less time than it took me to stake a tomato plant. To work in his garden was to commune with nature without ever leaving the marketplace.

By growing much more produce than he and Grandma could ever hope to consume, Grandpa transformed his vegetables into commodities. And to make sure of this elevated status, he planted exclusively those varieties sold by the supermarket chains: beefsteaks, iceberg lettuce, Blue Lake string beans, Marketmore cukes. Never mind that these were usually varieties distinguished less for their flavor than their fitness for transcontinental shipment; he preferred a (theoretically) marketable crop to a tasty one. Of course selling the vegetables wasn't a realistic option; he appreciated that an eighty-five-year-old real estate magnate couldn't very well open a farm stand, as much as he might have liked to. Still, he needed distribution channels, so he worked hard at giving the stuff away. All summer, before he got dressed for work (he never retired), Grandpa harvested the garden and loaded the trunk and backseat of his Lincoln with bushel baskets of produce. As he went on his rounds—visiting tenants, haggling with bankers and brokers, buying low and selling high—he'd give away baskets of vegetables. Now my grandfather never gave away anything that didn't have at least some slender string attached to it, and no doubt he believed that his sweet-as-sugar beefsteaks put these businessmen in his debt, gave him some slight edge. And probably this was so. At the least, the traveling produce show put the suits off their guard, making Grandpa seem more like some benign Old World bumpkin than the shark he really was.

It took a long time before I understood the satisfactions of giving away vegetables, but the pleasures of harvesting them I acquired immediately. A good visit to Grandma and Grandpa's was one on a day he hadn't already harvested. On these occasions I could barely wait for Grandpa to hand me a basket and dispatch me to the garden to start the picking. Alone was best—when Grandpa came along, he would invariably browbeat me about some fault in my technique, so I made sure to get out there before he finished small-talking with Mom. Ripe vegetables were magic to me. Unharvested, the garden bristled with possibility. I would quicken at the sight of a ripe tomato, sounding its redness from deep amidst the undifferentiated green. To lift a bean plant's hood of heart-shaped leaves and discover a clutch of long slender pods hanging underneath could make me catch my breath. Cradling the globe of a cantaloupe warmed in the sun, or pulling orange spears straight from his sandy soil—these were the keenest of pleasures, and even today in the garden they're accessible to me, dulled only slightly by familiarity.

At the time this pleasure had nothing to do with eating. I didn't like vegetables any better than most kids do (tomatoes I considered disgusting, acceptable only in the form of ketchup), yet there it was: the vegetable sublime. Probably I had absorbed my grandfather's reverence for produce, the sense that this was precious stuff and here it was, growing, for all purposes, on trees. I may have had no use for tomatoes and cucumbers, but the fact that adults did conferred value on them in my eyes. The vegetable garden in summer made an enchanted landscape, mined with hidden surprises, dabs of unexpected color and unlikely forms that my grandfather had taught me to regard as treasures. My favorite board game as a child was Candyland, in which throws of the dice advanced your man through a stupendous landscape of lollipop trees, milk-chocolate swamps, shrubs made of gumdrops. Candyland posited a version of nature that answered to a child's every wish—a landscape hospitable in the extreme, which is one definition of a garden—and my grandfather's vegetable patch in summer offered a fair copy of that paradise.

This was Grandpa's garden. If I could look at it and see Candyland, he probably saw Monopoly; in both our eyes, this was a landscape full of meaning, one that answered to wishes and somehow spoke in a human language. As a child I could always attend more closely to gardens than to forests, probably because forests contain so little of the human information that I craved then, and gardens so much. One of the things childhood is is a process of learning about the various paths that lead out of nature and into culture, and the garden contains many of these. I can't imagine a wilderness that would have had as much to say to me as Grandpa's garden did: the floral scents that intimated something about the ways of ladies as well as flowers, the peach tree that made legible the whole idea of fruit and seed, the vegetables that had so much to say about the getting of food and money, and the summer lawns that could not have better expressed the hospitality of nature to human habitation.

My parents' yard (you would not call it a garden) had a lot to say, too, but it wasn't until I was much older that I could appreciate this. Landscapes can carry a whole other set of meanings, having to do with social or even political questions, and these are usually beyond the ken of young children. My father's unmowed front lawn was a clear message to our neighbors and his father-in-law, but at the time I was too young to comprehend it fully. I understood our yard as a source of some friction between my parents, and I knew enough to be vaguely embarrassed by it. Conformity is something children seem to grasp almost instinctively, and the fact that our front yard was different from everybody else's made me feel our family was odd. I couldn't understand why my father couldn't be more like the other dads in the neighborhood.

One summer he let the lawn go altogether. The grasses grew tall enough to flower and set seed; the lawn rippled in the breeze like a flag. There was beauty here, I'm sure, but it was not visible in this context. Stuck in the middle of a row of tract houses on Long Island, the lawn said *turpitude* rather than *meadow,* even though that is strictly speaking what it had become. It also said, to the neighbors, *fuck you.*

A case could be made that the front lawn is the most characteristic institution of the American suburb, and my father's lack of respect for it probably expressed his general ambivalence about the suburban way of life. In the suburbs, the front lawn is, at least visually, a part of a collective landscape; while not exactly public land, it isn't entirely private either. In this it reflects one of the foundations of the suburban experiment, which Lewis Mumford once defined as "a collective effort to live a private life." The private part was simple enough: the suburban dream turns on the primacy of family life and private property; these being the two greatest goods in my father's moral universe, he was eager to sign up. But "owning your own home" turned out to be only half of it: a suburb is a place where you undertake to do this in concert with hundreds of other "like-minded" couples. Without reading the small print, my father had signed on for the whole middle-class utopian package, and there were heavy dues to pay.

The front lawn symbolized the collective face of suburbia, the backyard its private aspect. In the back, you could do pretty much whatever you wanted, but out front you had to take account of the community's wishes and its self-image. Fences and hedges were out of the question: they were considered antisocial, unmistakable symbols of alienation from the group. One lawn should flow unimpeded into another, obscuring the boundaries between homes and contributing to the sense of community. It was here in the front lawn that "like-mindedness" received its clearest expression. The conventional design of a suburban street is meant to forge the multitude of equal individual parcels of land into a single vista—a democratic landscape. To maintain your portion of this landscape was part of your civic duty. You voted each November, joined the PTA, and mowed the lawn every Saturday.

Of course the democratic system can cope with the nonvoter far more easily than the democratic landscape can cope with the nonmower. A single unmowed lawn ruins the whole effect, announcing to the world that all is not well here in utopia. My father couldn't have cared less. He owned the land; he could do whatever he wanted with it. As for the neighbors, he felt he owed them nothing. Ours was virtually the only Jewish family in a largely Catholic neighborhood, and with one or two exceptions, the neighbors had always treated us coolly. Why should he pretend to share their values? If they considered our lawn a dissent from the common will, that was a fair interpretation. And if it also happened to rankle his father-in-law, well, that only counted in its favor. (One should be careful, however, not to minimize the influence of laziness on my father's philosophy of lawn care.)

The summer he stopped mowing altogether, I felt the hot breath of a tyrannical majority for the first time. Nobody would say anything, but you heard it anyway: *Mow your lawn.* Cars would slow down as they drove by our house. Probably some of the drivers were merely curious: they saw the unmowed lawn and wondered if perhaps someone had left in a hurry, or died. But others drove by in a manner that was unmistakably expressive, slowing down as they drew near and then hitting the gas angrily as they passed—this was pithy driving, the sort of move that is second nature to a Klansman.

The message came by other media, too. George Hackett, our next-door neighbor and my father's only friend in the development, was charged by the neighbors with conveying the sense of the community to my father. George didn't necessarily hold with the majority on this question, but he was the only conceivable intermediary and he was susceptible to pressure. George was a small, somewhat timid man—he was probably the least intimidating adult in my world at the time—and I'm sure the others twisted his arm fairly hard before he agreed to do their bidding. It was early on a summer evening that he came by to deliver the message. I don't remember it all, but I can imagine him taking a drink from my mother, squeaking out what he had been deputized to say, and then waiting for my father—who next to George was a bear—to respond.

My father's reply could not have been more eloquent. He went to the garage and cranked up the rusty old Toro for the first time since spring; it is a miracle the thing started. He pushed it out to the curb and then started back across the lawn to the house, but not in a straight line; he swerved right, then left, then right again. He had made an *S* in the tall grass. Then he made an *M* and finally a *P.* These were his initials, and as soon as he finished writing them, he wheeled the lawn mower back to the garage, never to start it up again.

It wasn't long after this incident that we moved out of Farmingdale. The year was 1961, I was six, and my father was by now doing well enough to afford a house on the more affluent north shore, in a town called Woodbury. We bought one of the first houses in a new development called the Gates; the development was going in on the site of an old estate, and the builder had preserved the gigantic, wrought-iron entrance gates in order to lend the new neighborhood a bit of aristocratic tone.

To the builder goes the privilege of naming the streets in his development, and the common practice then was to follow a theme. Most neighborhoods had streets named for trees and flowers, but the Gates from the start pictured itself as a different kind of development—grander, more forward-looking—so it would have a different kind of street name. Alaska had recently been made the fiftieth state, and this developer, regarding himself perhaps as a pioneer or empire builder, decided to name all his streets after places there; our house was at the corner of Juneau Boulevard and Fairbanks Drive. (The word *street,* with its urban connotation, is not a part of the suburban vocabulary.) The incongruity of remote, frontierish place names attached to prissy "boulevards" and "drives" and "courts" never seemed to bother anybody.

With a new development, you chose your plot of land, one of the three available house types (ranch, colonial, or split-level), and then they built it for you. We chose a wooded acre (a vast tract compared to what we had in Farmingdale) that sloped down from Juneau Boulevard into a hollow. The topography afforded some privacy, but it meant that the floor of our basement was usually under several inches of water. As for house type, there could be no question: we always lived in ranch houses. There were two reasons for this. First, a ranch was the most "modern" kind of house, and my parents regarded themselves as modern. The second reason had to do with safety: my mother believed you simply did not raise children in a house with a staircase. You might as well invite the Long Island Rail Road to lay its tracks through your backyard.

After the contract had been signed, my father would drive my sister and me to Woodbury each weekend to follow the progress of our new house. We watched as the wooded acre was partially cleared and staked out by surveyors with tripods. My parents had chosen this plot because of its deep oak forest, and we tied ribbons to the trees we wanted saved, including a great big two-trunked oak that would stand outside our front door for the rest of my childhood. We felt like pioneers, watching as the woods gave way to bulldozers and a whole new landscape began to take shape. I remember being deeply impressed by what the heavy equipment could accomplish; who knew a forest could be turned into a yard, or a hill made to disappear? I'd never seen land change like this. The day they came to pour the foundation, my father gave us pennies to drop in the fresh concrete for good luck.

Though only twenty minutes away from Farmingdale, the Gates was a different world. Farmingdale was a blue-collar neighborhood, inhabited by electricians, engineers, and aerospace workers for whom a suburban home was the first and perhaps the only proof of membership in the American middle class. It may have been the tenuousness of our neighbors' grip on that identity that made them so touchy about lawns and Jews. The people who bought into the Gates, on the other hand, were the sons and daughters of the lower middle class, which in the fifties and sixties meant they were on their way to becoming quite affluent; they were lawyers and doctors and the owners of small businesses. This was a more confident class, and they sought a suburban home that would reflect their ascendancy and sophistication. Already in the early 1960s, the suburbs had acquired a reputation for conformity and squareness, and the Gates appealed to people who wanted to live in a suburb that didn't look like one. The streets were broad and, instead of being laid out in a tight grid, they curved in unpredictable ways. There was no practical reason for this, of course; the streets didn't curve *around* anything. They curved strictly to give an impression of ruralness and age. A sort of antisuburban suburban aesthetic ruled the development: the plots had been cut into irregular shapes, sidewalks had been eliminated, and roads ended in cul-de-sacs (these were the "courts").

Compared to Farmingdale, the landscaping in the Gates was wildly expressive. Not that the tyranny of the front lawn had been overturned. But even within that tight constraint, many families managed, in a phrase you were beginning to hear a lot, to do their own thing. Most of the landscaping styles were vaguely aristocratic, recalling the look of British country estates or, even more improbably, southern plantations. Circular driveways were very big. These broad crescents, scrupulously outlined in shrubbery, would curve right up to the front door. The planting served to emphasize the asphalt, which would be repainted each year with driveway sealer to restore its inky sheen. These driveways made a visitor feel he was driving up to a mansion rather than a split-level; you half expected someone in livery to open the car door for you. But the true purpose of the circular driveway was to provide a glittering setting for the family jewel, which was usually a Cadillac or Lincoln. Circular driveways make it socially acceptable to park your car right in the middle of the front yard where no one could possibly miss it.

The Rosenblums, a few doors up Juneau Boulevard from us, had *two* driveways, one on each side of the biggest, flattest, most pristine lawn in the development. Their aloof white colonial stood squarely in the middle of this vast green rectangle, framed by the two dead-straight black pavements. One driveway delivered family members to the garage and the other brought guests to a somewhat more formal entrance. The façade of the house was vaguely Greek Revival, but immense, with four ridiculous Doric columns and a giant wrought-iron chandelier hanging in the middle. It always reminded me of Tara. Just what kind of fantasy Mr. Rosenblum was working out here I have no idea, but I do remember he would get hopping mad whenever anyone used the wrong driveway.

It must have been obvious to my parents that the "S.M.P." approach to lawn care and gardening would not go over in the Gates. Fortunately, they could now afford to buy a fancy landscaping job and, even more important, a maintenance contract that would help keep my father on the right side of his new neighbors. It's important to understand that my parents were not indifferent to the landscape; even my father cared about his trees and shrubs. He simply didn't like lawns and preferred to deal with the rest of the garden at a remove, ideally through a window. But with money came a new approach to gardening, one that replaced laborious, direct involvement with the earth and plants with practices more to his liking: supervision, deal making, shopping, technological tinkering, negotiation. One must enlarge the definition of gardening a bit before his quasi-horticultural accomplishments can be fully appreciated. Perhaps the greatest of these involved the weeping birch that stood in the middle of our backyard in Farmingdale, forming what looked like a cascading green fountain. This somewhat rare specimen was my mother's favorite tree, and she wanted very badly to bring it with us to Woodbury. So as soon as the contract to sell the house in Farmingdale had been signed, but before the new owners moved in, my father arranged to have Walter Schikelhaus, my grandfather's landscape man, dig it out and truck it to Woodbury. But the tree was so distinctive, and occupied such a pivotal position in the backyard, that the new owners were bound to miss it. So my father had Walter plant in its place a weeping willow. Then he instructed Walter to paint the willow's bark white and carefully prune its branches to resemble a weeping birch. After mowing his initials in Farmingdale, this was perhaps my father's greatest achievement as a gardener, a strikingly original synthesis of topiary and fraud.

The man my parents hired to design, plant, and maintain our yard must have been a renegade among Long Island landscapers. Taking his cues from my father, he came up with a radical, low-maintenance design that included only a slender, curving ribbon of lawn. This narrow lane of sod wound an unpredictable path among every alternative to grass then known to landscaping: broad islands of shrubbery underplanted with pachysandra; flagstone patios; substantial wooded areas; and even a Japanese section paved in imported white pebbles. It was all very modern, and though it defied the conventions of suburban landscape design, it did so with taste. Overall, the front yard had far more ground cover than grass. Instead of foundation planting, most of the shrubs (rhododendron and azalea, in the main) were planted close to the street, forming a rough, irregular hedge that obscured the house. The retaining wall along the driveway was a terraced affair made out of railroad ties, which at the time were still a novelty in landscape design. (They weren't commercially available then, but my father arranged to buy them off trucks from LILCO and LIRR employees.) Much of the property was left wooded. And the Toro stayed behind, in Farmingdale. We may have been the only family on Long Island that didn't own a lawn mower.

Since my father's line on watering was more or less the same as his line on mowing, he decided to order a state-of-the-art sprinkler system. From his command post in the garage he would be able to monitor and water every corner of his acre, one zone at a time. An elaborate timer, working in conjunction with a device that judged the moisture content of the soil, was supposed to ensure that the grass and pachysandra enjoyed optimum conditions. But it soon became clear that the sprinkler man had taken my father for an expensive ride. We had hundreds more sprinkler heads than we could possibly need; every six feet another bronze mushroom poked out of the ground. And the system never worked properly.

Often in the middle of the night, or during a rainstorm, the sprinkler heads would suddenly start hissing and spitting in unison, as if under the direction of some alien intelligence. From some heads the spray roared like Niagara, but most of them dribbled pathetically. My father would spend hours at a time in the garage, standing in his boxer shorts at the control panel, trying vainly to rein in the system's perversity.

From my point of view, my father's remote-controlled landscape was sorely lacking. Once the crew finished planting the shrubs and laying down the carpet of sod, there was nothing left to do but look at it. For all its banality, the conventional suburban landscape, like the suburbs themselves, was tailored to the needs of children. As a place to play, nothing surpasses a lawn. Beautiful as it was, my parents' yard, with its sliver of lawn and masses of shade trees, was inhospitable to children; it was a spectator's landscape, its picturesque views best appreciated indoors, in boxers. You certainly couldn't play in the pachysandra.

But what it lacked most was a garden. True, considered whole, it *was* a garden, but to my mind (as in the common American usage) a garden was a small plot of flowers or vegetables; everything else was a "yard." A yard was just a place; a garden was somehow more specific and, best of all as far as I was concerned, it was productive: it *did* something. I wanted something more like my grandfather's garden, a place where I could put my hands on the land and make it do things. I'd also been spending a lot of time watching workmen revolutionize the landscape all around me as they created this new development: every day, it seemed, forests turned into lawns, fresh black roads bisected the nearby farm fields, sumps were being dug, whole hills were *moving.* Everywhere you looked the landscape seemed to be in flux, and I was taken with the whole idea of reshaping earth. Meanwhile our own acre had suddenly fossilized. All you could do was go to the garage and fiddle with the sprinkler controls. I wanted to *dig*.

Most of our yard now came under the jurisdiction of the maintenance crews that showed up every Friday, but there were still a few corners that escaped their attention. The lawn never took in the backyard, along the narrow corridor between the house and the woods; no matter what blend of seed they tried, the shade eventually defeated the grass. When the landscapers finally gave up on this patch I was allowed to dig in it. Of course the shade precluded my planting a garden, but I had another idea: to give the property a badly needed body of water. I ran a hose underground from the house and constructed a watercourse: a streambed lined with stones that passed through a complex network of pools and culminated in a spectacular waterfall, at least eight inches tall. I spent whole afternoons observing the water as it inscribed new paths in the ground on its infinitely variable yet inevitable journey toward the woods. I was learning to think like water, a knack that would serve me well in the garden later on. I experimented with various stones to produce different sounds and motions, and no doubt wasted an obscene amount of water. Though I judged it a miniature landscape of extraordinary beauty, my water garden may have really been little more than a mud patch; I'm not sure.

When I tired of my water garden, I ripped it out and built a cemetery in its place. We had lots of pets, and they were constantly dying. Not just cats and dogs, but canaries and chicks, turtles and ducklings, gerbils and hamsters. Whenever one of these animals expired, my sisters and I would organize elaborate funerals. And if all our pets happened to be in good health, there were always roadkills in need of decent burial. After we interred the shoebox-caskets, we would rake and reseed the ground and plant another homemade wooden cross above the grave. I understood that crosses were for Christians. But a Star of David was beyond my carpentry skills, and anyway I was inclined to think of pets as gentiles. To a child growing up Jewish, the Other, in all its forms, was presumed to be Christian.

My usual partner in all these various landscaping endeavors was Jimmy Brancato, an uncannily hapless boy who lived down the street with his problematic parents. Mr. Brancato was a vaguely gangsterish character who owned a car wash in Hempstead, and who, it was rumored, had once spent time in jail in another state. Mrs. Brancato, who wore her bleach-blond hair in a monumental do and looked a lot like a gun moll, was a champion screamer and worrier. She was so steadfast in the conviction that her children were destined for trouble (jail in Jimmy's case; out-of-wedlock pregnancy in his sisters') that they must have gradually come to believe there could be no alternative. And sure enough, one of the daughters eventually did get knocked up and Jimmy had a serious run-in with the law.

But that came much later; at the time I'm writing about, Jimmy was nine or ten, merely on the cusp of delinquency. As you can imagine, we both preferred to hang out at my house. Jimmy loved my mother, probably for the simple reason that she didn't see prison stripes when she looked at him. And I was too terrified of Jimmy's parents to go near them voluntarily. I liked Jimmy because, compared to me, he was bold and fearless; he liked me because, compared to him, I had a brain. We made a good team.

We both liked to garden, though it's possible Jimmy was just following my lead here. I usually set the agenda, explaining to Jimmy where we were going to dig or what we were going to plant that day, citing my grandfather whenever I needed to bolster my authority. Our first garden, which we called a farm, was terraced: the railroad-tie retaining wall rose from the driveway in a series of four or five steps, each of which made a perfect garden bed. We'd plant strawberries on one level, watermelons on another, and on a third some cucumbers, eggplants, and peppers. But strawberries were by far our favorite crop. They had the drama of tomatoes (the brilliant red fruit), they came back every year by themselves (something we thought was very cool), *and* they were edible. Our goal, though, was to harvest enough strawberries to sell—this being a farm—and anytime we could get six or seven ripe ones at a time, we'd put them in a Dixie cup and sell them to my mother. Eventually we hoped to open a farm stand on Juneau Boulevard. Jimmy always worked like a dog. Even after I'd be called in for dinner, he'd stay out there digging and hoeing until his mother stuck her head out of their kitchen window and started hollering for him to come home.

As much as he seemed to enjoy it, this form of gardening didn't fully satisfy Jimmy's taste for adventure; perhaps he sensed that it would be hard to realize his destiny in the vegetable patch (though in fact he eventually would find a way to do exactly that). Jimmy held a relatively broad concept of gardening, embracing as it did such unconventional practices as the harvesting of other people's crops in their absence. Bordering our development was a pumpkin field, and several times each fall Jimmy insisted I accompany him on a mission to steal as many pumpkins as we could pile in our wagons. Going along was the price I paid for Jimmy's help on the farm.

The pumpkin field in October was a weirdly beautiful place, with its vast web of green vines blanketing the gorgeous orange orbs for as far as you could see. Here was the vegetable sublime again, but now its experience was fraught with danger. I'd been taught that trespassing was a heinous crime, and the NO TRESPASSING sign we had to drag our wagons past choked me with fear. In the suburbs private property was such a sacrosanct institution that even young children felt its force. Jimmy claimed—probably just to scare me but you never knew for sure—that the farmers had rifles that fired bullets made of salt, and if they saw us they would be fully within their rights to shoot since *we were on their property.* These salt pellets were said to cause excruciating pain. (As if getting shot with steel bullets wouldn't have been bad enough.) We managed to get out alive every time, but I have to say I wasn't entirely disappointed the year the pumpkin field gave way to a new housing development.

After we'd arrived safely at home with our pumpkins (we'd always go to Jimmy's; my mother would have flipped out if we'd shown up with hot pumpkins), we'd divvy up the loot and then Jimmy would proceed methodically to smash his share. This was a pleasure I could not comprehend. But clearly the kick for Jimmy came in stealing the pumpkins, not owning them. Watching him get off bashing his pumpkins, you would think he'd been possessed. And the longer I knew him, the more I began to sense that he had an almost mystical attraction to trouble. One summer while my family was away on vacation, Jimmy was running some routine experiments with matches when he accidentally burned down most of the forest behind our house. All kids chucked snowballs at passing cars, but when Jimmy did it he would smash a windshield and then actually get caught. He wasn't a bad kid, not at all; it's just that he had some sort of tropism that bent him toward disaster as naturally as a plant bends toward sunlight.

Years after we had gone our separate ways, Jimmy figured out a way to combine what I'd taught him about gardening with his penchant for trouble. It must have been around 1970, when he was in the ninth grade, that Jimmy decided to start his own farm, one that might actually make some money. He planted a small field of marijuana. Jimmy had considered all the angles and went to great lengths to avoid detection. Growing pot on his parents' property was obviously out of the question, so he cleared a plot down by the Manor House, the abandoned mansion on whose grounds the Gates had been built. The developer had promised to turn the Manor House into a community center, but he had skipped town long ago and the place had devolved into a kind of no-man's-land, a gothic ruin surrounded by old refrigerators and derelict shopping carts. Brambles and sumac choked any spot not occupied by a stripped Impala, and clearing a patch for a garden must have been back-breaking work. Most of us didn't dare go near the Manor House during the day, let alone after dark. But each night, after midnight, Jimmy would slip out of his house, ride his bicycle down to the Manor House, and tend his precious crop by flashlight.

Getting caught wasn't going to be easy, but Jimmy managed to pull it off.

Shortly before Jimmy planned to begin his harvest, a neighborhood boy riding his bicycle around the Manor House happened upon his garden. Today, the leaf pattern and silhouette of a pot plant is as familiar as a maple's, but this was not yet the case in 1970. Unfortunately for Jimmy, this particular boy had recently attended an assembly at school where a policeman had shown the kids how to recognize marijuana. The boy raced home and told his mother what he'd seen and his mother called the police.

Jimmy had by now been in enough scrapes to be well known to the local police and I'm sure they immediately settled on him as a prime suspect. In the version of the story I heard, when the cops dropped by to question Jimmy and his mother, he kept cool, admitting nothing. Since they had no evidence linking Jimmy to the marijuana plants, that should have been the end of it. But the police had aroused Mrs. Brancato's suspicions and she decided to conduct a search of Jimmy's room.

Of the seven deadly sins, surely it is pride that most commonly afflicts the gardener. Jimmy was justly proud of his garden, and though he knew better than to invite anyone to visit it, he apparently couldn't resist taking a few snapshots of his eight-foot beauties in their prime. Mrs. Brancato found the incriminating photographs and, concluding it would be best for her son in the long run, turned them in to the police. No charges were brought, but Jimmy was packed off to military school, and I lost track of him.

My own gardening career remained well within the bounds of the law, if not always of propriety. Around the same time Jimmy was tending his plot down at the Manor House, I moved the farm from the cramped quarters of the retaining wall to a more spacious plot I had cajoled from my parents alongside the foundation of our house. This would be my last garden in the Gates. Even the most devoted young gardener will find that his interest fades around the time of high school, and soon mine did. But the summer before I got my driver's license I made my most ambitious garden yet. I persuaded my parents to buy me a few yards of topsoil, and in the space of a hundred square feet I crammed a dozen different crops: tomatoes (just then become edible), peppers, eggplants, strawberries, corn, squash, melons (watermelon and cantaloupe), string beans, peas. Everything but lettuce, which, since it bore no fruit, held not nearly enough drama for me. Why would anyone ever want to grow *leaves*?

Years later when I read about European techniques of intensive agriculture, I realized this is what I had been doing without knowing it. I enriched the soil with bags of peat moss and manure, tilled it deeply, and then planted my seedlings virtually cheek-by-jowl. Since the bed was long and narrow, I decided to dispense with rows and planted most of the seedlings no more than six inches apart, in a pattern you would have to call free-form. Everything thrived: by August, my postage-stamp garden, haphazard though it was, was yielding bushels of produce.

Even my parents took note of this garden, marveling at the peppers and tomatoes I brought to the dinner table. But the person I really wanted to impress was my grandfather. By this point, my relationship with Grandpa was badly frayed. I wore my hair long and had grown a beard, and this deeply troubled him. By the time I turned fifteen, I could do nothing right by him, and visits to Babylon, which had held some of the sweetest hours of my childhood, had become an ordeal. From the moment I arrived, he would berate me about the beard, my studiously sloppy clothes, the braided leather bracelet I wore, and any other shred of evidence that I had become one of those despised *'ippies,* as he used to spit out the word. I figured that if there was one place where an elderly reactionary and an aspiring hippie could find a bit of common ground, it was in the vegetable garden. I had finally made a garden he'd be proud of, and when he and Grandma made one of their infrequent visits to our house that summer, I couldn't wait to take him around back and show him what I'd achieved.

But Grandpa never even saw the garden I had made. All he saw were weeds and disorder. You call this a garden? he barked. It's all too close together—your plants are going to choke each other out. And where are your rows? *There have to be rows.* This isn't a vegetable garden—what you've got here is a weed garden! The big red beefsteaks, the boxy green peppers, the watermelons now bigger than footballs: everything was invisible to him but the weeds. He looked at my garden and saw in it everything about me—indeed, everything about America in 1970—that he could not stand. He saw the collapse of order, disrespect for authority, laziness, the unchecked march of disreputable elements. He was acting like a jerk, it's true, but he was my grandfather, an old man in a bad time to be old, and when he got down on his knees and started furiously pulling weeds, I did feel ashamed.

So I guess you could say that Jimmy and I were expelled from our gardens at around the same time. But that would be too neat. For all I know, Jimmy today tends twenty acres of the finest sinse in Humboldt County. In my case, the arrival of my driver's license did more to push me out of the garden than my grandfather's intemperate attack on my technique. If gardening is an exploration of a place close to home, being a teenager is an exploration of mobility, and these two approaches to place, or home, are bound sooner or later to come into conflict. For at least a decade I probably didn't think once about plants or even notice a landscape. Eventually, though, I came back to the garden, which is probably how it usually goes. Much of gardening is a return, an effort at recovering remembered landscapes. I was lucky that when I took up gardening again my grandfather was still alive. He was over ninety by the time I had my own house, and he never did get to see it. But I would bring him pictures, carefully culled to give an impression of neatness and order, and, after examining them closely for evidence of weeds, he would pronounce his approval. By then, his own garden consisted of a half-dozen tomatoes planted by the back door of a small condominium. I would help him weed and harvest; he still grew enough beefsteaks to give a few away. He would ask me to describe my garden, and I would, choosing my words with care, painting a picture of a place that he would find hospitable. The garden I described was largely imaginary, combining elements of my actual garden with memories of Babylon and the kind of pictures that I suppose are common to every gardener's dreams. It was one of those places that is neither exactly in the past nor in the future, but that anyone who gardens is ever moving toward. It was somewhere we could still travel to together. On one of my last visits to see him, he told me I could have his Dutch hoe, declaring it was the best tool for weeding he had ever found. Grandpa was ninety-six, three times my age exactly, and though his step by then was uncertain, he took me outside and showed me how to use it.

Spring

CHAPTER 2

Nature Abhors a Garden

When I finally did come back to the garden, I was coming from the city and brought many of the city man's easy ideas about the landscape and its inhabitants. One of these had to do with the problem of pests in the garden, about which I carried the usual set of liberal views. To nuke a garden with insecticide, to level a rifle sight at the back of a woodchuck in flatfooted retreat, to erect an electric barricade around a vegetable patch: such measures, I felt, were excessive, even irresponsible. I took nature's fragility for granted, and the idea of crushing local opposition to my plans for the land simply by dint of superior firepower seemed reckless and unjust, an act of environmental imperialism. Besides, wildlife was one of the attractions of the country; the deer, foxes, porcupines, and woodchucks were what told you you were there. These animals had arrived long before the gardener, so who was the interloper here? And what was gardening about if not working out a more harmonious relationship with nature?

One of gardening's virtues is to clear the mind of easy sentiments about nature in general, and its fauna in particular. The first challenge to one's romance of animals comes in April, after you've broken your back turning the soil, humped bales of peat moss and bags of manure from the car trunk to the garden, dug these in by pitchfork, and then laid out in scrupulous rows the seedlings of early crops—lettuce, broccoli, cabbage. Do all that, kill an afternoon, and see how you feel the next morning when this orderly parade ground of seedlings has been mowed down by a woodchuck out snacking.

First you will feel frustration, at the waste of time, effort, and cash. Then a sense of persecution: with all the millions of tender shoots pushing up across the countryside this time of year, why have these animals chosen to dine at this particular plot? Now consider the forlorn appearance of the mowed-down rows, each seedling neatly snipped off a half inch above the ground, as if by someone with a pair of scissors and all the time in the world. This is what indicates a woodchuck is responsible. They devour a crop systematically, whereas a doe—nervous, and possessing perhaps a more developed sense of shame—will nibble a plant here, snip a shoot there, and then, startled by a falling leaf or something equally perilous to a two-hundred-pound mammal, dash off before her meal is done. The woodchuck, meanwhile, approaches the garden as if it were a spread laid out expressly for him; he regards your plants less as a thief might than a relative. He does not worry that his repast could be interrupted, and he fully intends to return tomorrow for seconds.

And the gardener will oblige, immediately replanting the mowed-down rows. For he is not about to fold his garden in the face of this lower-order impertinence. A rodent whose brain could fit in a thimble might win a battle or two, but finally the war must go to the larger brain. All of natural history is on the gardener's side. What is our species doing on this planet if not winning precisely this kind of contest?

At least that's how I saw matters the first time I woke to the evidence of a predawn April raid on my freshly planted vegetable garden. I assessed the damage, sized up my adversary, and decided that the wisest course of action was to take the battle to the woodchuck's own territory. I went looking for his burrow.

My vegetable garden is laid out on a small, flat lawn that ends, to the north, at the base of a small slope. The slope is overgrown with vetch, a tangle of blackberry briars, and a couple of Russian olive bushes—perfect cover, in other words, for a woodchuck burrow, and not five chuck-sized paces from the nearest garden row. Woodchucks, being both nearsighted and slow of foot, prefer to set up house as close to their favorite dining spot as prudence will allow. I whacked at the brush with a machete and there it was: a large ugly mouth set into the hillside with a pile of freshly dug soil arranged beneath it like a fat bottom lip. The woodchuck was not only visiting my garden, he had moved in for the season.

This called for a program of behavior modification. I gathered a half dozen fist-size rocks and squeezed them into the hole. Then I mounded a few shovelfuls of earth on top and stomped on it a few times to jam the rock and earth as far down into the tunnel as possible. This ought to persuade him to dine elsewhere, I decided, with all the confidence of someone who understood not the first thing about woodchucks.

The very next day the hole had yawned open and spit out the rocks and the soil. No doubt hungry from the work of excavation, the woodchuck had emerged from his burrow to sample a fresh planting of lettuce seedlings.

The reader might reasonably wonder at this point why it was that I had no fence. I was asked this question several times after the woodchuck struck and never came up with an entirely satisfactory answer. I could offer a few trivial explanations, having to do with economy and competence. But I suspect my reluctance to put up a fence was a more visceral matter. Fences just didn't accord with my view of gardening. A garden should be continuous with the natural landscape, I felt, in harmony with its surroundings. The idea that a garden might actually require *protection* from nature seemed absurd. Somewhere along the line I had been convinced that a fence bespoke disharmony, even alienation, from nature.

I suspect I had also absorbed the traditional American view that fences were Old World, out of place in the American landscape. This notion turns up repeatedly in nineteenth-century American writing about the landscape. One author after another denounces "the Englishman's insultingly inhospitable brick wall, topped with broken bottles." Frank J. Scott, an early landscape architect who had a large impact on the look of America's first suburbs, worked tirelessly to rid the landscape of fences, which he derided as a feudal holdover from Britain. Writing in 1870, he held that "to narrow our own or our neighbor's views of the free graces of Nature" was selfish and undemocratic. To drive through virtually any American suburb today, where every lawn steps right up to the street in a gesture of openness and welcome, is to see how completely such views have triumphed. After a visit to the United States, Vita Sackville-West decided that "Americans must be far more brotherly-hearted than we are, for they do not seem to mind being over-looked. They have no sense of private enclosure."

In a typical American suburb such as the one where I grew up, a fence or hedge along the street meant one thing: the family who lived behind it was antisocial, perhaps even had something to hide. Fences and hedges said: Ogres within; skip this place on Halloween. Except for these few dubious addresses, each little plot in our development was landscaped like a miniature estate, the puniest "expanse" of unhedged lawn made to look like a public park. I don't know about "brotherly-heartedness," though. Any enjoyment of this space was sacrificed to the conceit of wide-open land, for without a fence or hedge, front yards were much too public to spend time in. Families crammed their activities into microscopic backyards, the one place where the usefulness of fences and hedges seemed to outweigh their undemocratic connotations.

But the American prejudice against fences predates the suburban development. Fences have always seemed to us somehow un-American. Europeans built walled gardens; Americans from the start distrusted the *hortus conclusus.* If the space within the wall was a garden, then what was that outside the wall? To the Puritans the whole American landscape was a promised land, a sacred space, and to draw lines around sections of it was to throw this paramount idea into question. When Anne Bradstreet, the Massachusetts colony's first poet, set about writing a traditional English garden ode, she tore down the conventional garden wall—or (it comes to the same thing) made it capacious enough to take in the whole of America. The Puritans had not crossed the Atlantic to redeem some small, walled plot of land; that they could have done in England. They, or rather God acting through them, had plans for all of it.

The transcendentalists, too, considered the American landscape "God's second book" and they taught us to read it for moral instruction. Residues of this idea persist, of course; we still regard and write about nature with high moral purpose (and, almost as often as it did in the nineteenth century, the approach produces a great deal of pious prose). And though in our own nature writing guilt seems to have taken the rhetorical place of nineteenth-century ecstasy, the essential religiosity remains. We may no longer spell it out, but most of us still believe the landscape is somehow sacred, and to meddle with it sacrilegious. And to set up hierarchies within it—to set off a garden from the surrounding countryside—well, that makes no sense at all.

Once you accept the landscape as a moral and spiritual space, ornamental gardening becomes problematic. For how can one presume to remake God's landscape? It is one thing to cultivate the earth for our sustenance—the Bible speaks of that—but to do so for aesthetic reasons has until very recently struck Americans as frivolous, or worse. Allen Lacy reports that, in combing American garden writing for his recent anthology (*The American Gardener,* which is the source of many of the historical quotations in this chapter), he found no discussion before 1894 of color or fragrance. We gardened for a variety of reasons—moral, spiritual, therapeutic, and economic—but aesthetic pleasure was not one of them. Even when we make pleasure gardens today, we do our best to hide the hand of the artist, avoiding anything that looks designed or artificial. We favor gardens that resemble natural landscapes, and that leaves little room for fences.

Long before I had read much about American approaches to the landscape, I unwittingly made a perennial border beholden to these ideas. The border runs beneath an old stone retaining wall along a narrow lawn that draws the eye away from the house and back toward a small pasture and a wood beyond that. I tried to design the border so that it wouldn't have a distinct beginning or end. As it moves back in space, the plants get rougher and bigger. The aristocratic refinements of delphinium, baby's breath, campanula, and lady's mantle gradually give way to day lilies, a sloppy drift of evening primrose, an ill-mannered six-foot-tall clump of rudbeckia and, finally, to proletarian purple loosestrife, a weedy plant that grows wild around here. From the house you cannot pinpoint where the border ends and the natural landscape resumes. If I wanted to put a fence around this garden, where would it go? A fence could only wreck this garden.

But so, it turns out, can woodchucks, deer, and meadow grasses. My early efforts at harmonious design were lost on the surrounding landscape, whose inhabitants promptly sought to take advantage of my naïve romanticism. The deer relished the young shoots of day lilies and delphinium. Woodchucks judged the loosestrife an ideal cover for a burrow exit. And the grasses from the meadow soon exposed the so-called hardy perennials as pushovers. Instead of the flower border pushing back toward the meadow, the meadow pushed forward toward the house and it met scant local resistance. Without my intervention, the border would not have survived its first season.

Under this many-fronted assault, it did not take long for most of my easy, liberal attitudes toward the landscape to fall. I soon came to understand the distance between the naturalist, who gazes benignly on all of nature's operations, and the experienced gardener, who perforce has developed a somewhat less sentimental view. Particularly toward woodchucks. I am not ready to see them banished from the planet altogether—surely they serve *some* ecological purpose—but I seriously doubt that news of some form of woodchuck megadeath in this part of the country would put me in an elegiac frame of mind.

But in giving up my romantic views of the local fauna, I may have gone overboard in the opposite direction. I tried everything I could think of to eliminate my woodchuck, in an escalating series of measures only a William Westmoreland could have completely understood. I started out with elaborately planned campaigns of behavior modification—my "send in a few advisers" phase, in which I confidently deployed the accumulated wisdom of Western civilization. I had done my research and discovered that woodchucks were scrupulous about personal hygiene. They set aside a room in their burrow to serve as a latrine. And they hate to dirty the fur on their bellies. Confident I had located my adversary's Achilles' heel, I introduced a few carefully selected substances into his tunnel: a dozen eggs, smashed and dribbled down its sides. A pint jar of molasses. Half a can of motor oil. A dead field mouse. And, lastly, a quart of creosote, vile stuff so sticky the woodchuck would need to have the fur on his belly steam-cleaned.

When this didn't work—evidently my woodchuck lacked his species' Felix Unger gene—I found myself attracted to less cerebral approaches. It's astonishing, actually, how much anger an animal's assault on your garden can incite. It was not as if I were liable to go hungry as a result of his depredations, after all. No, this was no longer merely a question of vegetables or even self-interest. This was about winning.

A rifle was out of the question; I've always been terrified of guns, and have never owned one. But I came up with something equally unsentimental: I found a somewhat flattened woodchuck along the highway, scooped it into a crate and brought it home. Then I jammed the carcass as far into the burrow as it would go. This was an act of terrorism, I admit. But either my woodchuck did not grasp its significance, or he chose to disregard it, because in two days' time, he had dug a detour around the corpse and the pillaging resumed.

I decided now to incinerate the woodchuck in his burrow. I had seen an item on the news concerning cabin fires aboard jetliners. In order to test a new, supposedly less combustible fuel, the FAA had simulated a cabin fire, and the footage they showed of fire racing wildly through the narrow enclosed space gave me an idea of exactly the sort of end the woodchuck deserved.

Take a moment to picture it.

So I poured maybe a gallon of gasoline down the burrow, waited a few minutes for it to fan out along the various passageways, and lit a match.

Evidently there was not much oxygen down there, because the flames shot in the wrong direction, up toward my face. I leapt back before I was singed too badly, and watched a black-orange fountain of flame flare out from the earth and reach for the overhanging olive bush. I managed to smother the fire with earth before the entire garden went up.

I guess this was my destroy-the-village-in-order-to-save-it phase.

Well, if fences are out of place in the American garden, where exactly do gasoline fires fit in? Fortunately, my brush with general conflagration among the vegetables shocked me out of my Vietnam approach to garden pests before I'd had a chance to defoliate my property or poison the ground water. But my fury at the woodchuck put me in touch with a few of our darker attitudes toward nature: the way her intransigence can make us crazy, and how willing we are to poison her in the single-minded pursuit of some short-term objective. You think you know better until you've been beset by cabbage worms or aphids and then seen just how fast a shot of some state-of-the-art petrochemical can wipe them out. But after the firefight I resolved to keep my head and think more in terms of containment than victory.

I also began to see that there might be more going on here than a cartoonish war between me and a woodchuck: big creature thwarted again and again by wily little creature; numerous laughs at big creature's expense. The cartoon was part of the story, but not all of it.

I realized this during a long walk one spring afternoon in the woods near my house. Most of the land around here is postagricultural hardwood forest; the farms were abandoned starting around the turn of the century, and the forest has made quick work of reclaiming large parts of the countryside. You might think the oak forest was primordial if not for the stone walls and other lingering signs of onetime cultivation: wolf trees (specimens with broad crowns, indicating they matured in open, uncompetitive spaces); the conspicuous blossom of a leggy old apple tree in May; even faint plow furrows still visible in the snow cover. But on this particular walk I found an even ghostlier set of signs. Following an old logging trail, I came to an area that seemed somehow more ordered than the surrounding woods. On both sides of the trail stood stone walls—linear piles, really—marking small rectangular enclosures among the trees. Within each square was a rectangular pit lined with rocks: the foundation of a small house.

I had stumbled upon Dudleytown, an abandoned nineteenth-century settlement that I'd often heard was nearby but had never been able to locate. Traces of former habitation were everywhere, like shadows on the landscape, even though the forest had completely colonized the area. Oaks, hickories, ash, and sycamores had spread out evenly over the village like a blanket, rising up in the former yards and fields and even in the middle of the cellar pits, jutting heedlessly through spaces that once had been organized into kitchens and bedrooms, warm spaces that had vibrated with human sounds.

If you blotted the trees from your sight and followed the contours of the land, you could make out the organization of the village. Houses lined a main street. The stone walls delineated each family's yard; in some stood gnarled apple trees on their last legs, starved for sunlight by the new forest canopy. Lilacs and clumps of day lily survived here and there, along with deep green patches of myrtle: the remnants of dooryard gardens that the forest had failed to vanquish. Some yards opened onto what must have been fields or pastures. Stone walls that had once marked legal boundaries and kept cows from straying threaded arbitrary paths through the trees, accomplishing nothing.

It is a spooky place. I'm not talking only about the ghostliness of abandoned settlement, or the weight of the past one often feels among ruins. What makes Dudleytown spooky is the evident speed and force and thoroughness with which the forest has obliterated the place. In the space of a few decades it has erased virtually every human mark.

To the gardener in me, Dudleytown assumed a spectral presence. Every weed I pulled, every blade of grass I mowed, each beetle I crushed—all now was done to slow its advance. Dudleytown made me see that the woodchuck was no free-agent pest, snacking strictly on his own account. He was part of a larger, more insidious threat: he labored on behalf of the advancing forest. Not only the animals, but the insects, the weeds, even the fungi and bacteria, were working together to erase my garden—and after that my lawn, my driveway, my patio, even my house. Does this sound a little paranoid? Perhaps it is, but my experience in the garden has taught me that nature seems to resent our presence here. She deploys a variety of agents, different depending on where you live, to undo our work in the garden. To what end? That also depends on where you live, but now I knew her local aim: Dudleytown.

The forest, I now understood, is "normal"; everything else—the fields and meadows, the lawns and pavements and, most spectacularly, the gardens—is a disturbance, a kind of ecological vacuum which nature will not abide for long. If it sometimes seems as if she has singled out the garden for special attention, that's because the "vacuum" here is greatest. Here the soil is richest and most frequently disturbed: what softer, sweeter, more hospitable bed could an airborne weed seed ever find to lie down in? The annual weeds, first to colonize a neglected garden, come this way, around here mostly ragweed, pigweed, touch-me-not, and smartweed. But the perennial weeds—the goldenrod, pokeweed, milkweed, and bindweed—can creep into your beds otherwise, often dispatching rhizomes underground, sometimes as far as fifty feet, in search of sweet soil. Others don't even have to find your garden: thousands of weed seeds lie dormant in every cubic foot of soil, patiently waiting for just the right combination of light and moisture before setting on your plants.

And garden plants are sitting ducks. Just as cultivated soil constitutes a kind of vacuum in the environment, so do most of the plants we choose to grow in it. What distinguishes cultivated fruits and vegetables is that they contain carbohydrates, proteins, and fats in greater concentrations than most wild plants. They stick out in the natural landscape like rich kids in a tough neighborhood. This is where the animals come in. The woodchucks, deer, and raccoons are the flora's great levelers, making sure there are no undue concentrations of nutritional wealth in the landscape; they'd consider themselves democrats if they considered at all. They want to redistribute my protein. But if their politics appeals to your egalitarianism, keep in mind that their tactics are not those of the social democrat.

Should the vertebrates fail to intimidate me into ceding my garden to the forest, a dozen different insect species, each with its own distinctive preferences, tactics, and disguises, will march on my plants in a series of waves beginning in April and not relenting till frost. First the cutworms, who saw off the seedlings at ground level. Then the aphids, specs of pale green that cluster on the undersides of leaves, sucking the vital fluids from young plants until they turn a last-gasp yellow. Next come the loathsome slugs: naked bullets of flesh—evicted snails—that hide from the light of day, emerging at sunset to cruise the garden along their own avenues of slime. The cabbage loopers are the paratroopers of the vegetable patch: their eggs are dropped on the cole crops by troop transports disguised as innocuous white butterflies. Last to arrive is the vast and far-flung beetle family—Colorado potato beetles, blister beetles, flea beetles, bean leaf beetles, cucumber beetles, Japanese beetles, Mexican bean beetles—who mount a massive airborne invasion beginning in midsummer.

Like the vertebrates, this exoskeletal mob is drawn by the nutritional extravagance of the vegetable garden, as well as by the fact that most garden plants are, let's face it, nature's weaklings. We breed garden plants primarily for qualities that appeal to us, not ones that might help ensure their survival. And the characteristics that most appeal to us—doubleness in flower blossoms, slowness to bolt in lettuce—are about as helpful in battle as designer fatigues. ("Disease resistance" is an afterthought, and usually a case of too little, too late.) Rather than school them in the martial arts, we enter into a tacit pact with our plants: in exchange for their beauty and utility, we shield them from the horrors of Darwinian struggle.

So don't lecture me about harmony in the garden. Or about the continuity of gardens and the natural landscape. The forest is so vigorous around here, and so well served by its advance guard of animals and bugs and weeds, that a single season of neglect would blast my garden back to meadow, and a decade would find the forest licking at my front stoop, while that dark conspiracy of microorganisms we call rot goes to work on the house itself. In fifty years: Dudleytown. A cellar pit with a sycamore rising through it.

What was the right approach to pests in the garden? How could I halt the advance of Dudleytown without turning my garden into a toxic waste site? I was beginning now to see that these questions quickly led to larger ones about how we choose to confront the natural landscape. Domination or acquiescence? As developers or naturalists? I no longer think the choice is so obvious.

Domination, translated into suburban or rural terms, means lawn. A few acres of Kentucky bluegrass arranged in a buffer zone between house and landscape, a no-man's-land patrolled weekly with a rotary blade. The lawn holds great appeal, especially to Americans. It looks sort of natural—it's green; it grows—but in fact it represents a subjugation of the forest as utter and complete as a parking lot. Every species is forcibly excluded from the landscape but one, and this is forbidden to grow longer than the owner's little finger. A lawn is nature under totalitarian rule.

On the other side is acquiescence: the benign gaze of the naturalist. Certainly his ethic sounds nice and responsible, but have you ever noticed that the naturalist never tells you where he lives? Unless you live in the city or a tent, the benign gaze is totally impractical—sooner or later it leads to Dudleytown.

The trick, I realize now, is somehow to find a middle ground between these two positions. And that is what a garden is, or should be: a midspace between Dudleytown and the parking lot, a place that admits of both nature and human habitation. But it is not, as I had imagined, a harmonious compromise between the two, nor is it stable; from what I can see, it requires continual human intervention or else it will collapse. The question for the gardener—and in a way it's a question for all of us—is, What is the proper character of that intervention?

Even my limited experience in the garden suggests that finding a good answer to that question will involve a much more complicated set of choices than the usual American alternatives, which seem to consist of either raping the land or sealing it away in a preserve where no one can touch it. That the first approach is bankrupt goes without saying. Yet, right as it sounds, the second one may be a dead end too. Gardening quickly teaches you to distrust all such absolutes, to frame the questions a little differently. Must we *always* shrink before our own power in nature? We are one of only a handful of creatures with the capacity to deliberately alter our environment. To simply renounce that power—isn't that in some sense to renounce our humanity? *Our* nature? And is that nature any less real than the nature we seem to think exists only *out there?* The poet and critic Frederick Turner, in a *Harper's Magazine* essay that seeks to break us of our habit of seeing nature and culture as opposed, asks why it is we can't see ourselves, and what we make and do, as part and parcel of nature. He cites the reply of Shakespeare's Polixenes, in *The Winter's Tale,* to Perdita, who spurns the hybridized flower because it is "unnatural": "This is an art/Which does mend Nature—change it rather; but/The art itself is nature."

For the gardener, breaking free of the notion that art always negates nature is liberating. Fresh aesthetic prospects open up, of course, but more to the point, a promising strategy against pests can begin to take shape. For starters, one can now reexamine the American taboo against fences. Fences may offend American ideas about democracy, limitlessness, and the landscape's sanctity, but perhaps we need to consider the possibility that their absence offends the idea of a garden. For most of history people have been making gardens and most of their gardens have been walled or fenced. The word *garden* derives from the old German word for enclosure, and the *O.E.D.*'s definition begins, "An enclosed piece of ground. . . ." (Compare that to *Webster's,* which makes no mention of the idea of enclosure.) Writing in 1914, George Washington Cable pointed out that "a gard, yard, garth, garden, used to mean an enclosure, a close, and implied a privacy to its owner superior to any he enjoyed outside of it. . . . Our public spirit and our imperturbability are flattered by [fencelessness], but our gardens . . . have become American by ceasing to be gardens." The long history of gardens, which traverses so many very different cultures, suggests that perhaps there is something natural about erecting a wall against the landscape on one side and society's gaze on the other. We number the beaver dam among nature's creations; why not also the garden wall?

The time had come for me to put up a fence. I went with five feet of galvanized steel mesh stretched across posts that had been treated with arsenic to resist rot and then sunk three feet into the earth. The bottom edge of the fence runs a foot underground, to deter the tunnelers. It doesn't look at all bad, and even though the wire mesh is invisible at a distance, when I close the garden gate behind me I feel as though I've entered a privileged space.

But much more important is the fact that, so far, the woodchuck respects the fence; the cabbages have reached softball size unmolested. The woodchuck doesn't appear to have abandoned his burrow, however, and I picture him jealously pacing the garden perimeter at dawn, scheming, looking for an angle. I remain on alert.

Now four feet of fence won't impede a doe with snap beans on her mind, but I can take care of her, too. Six inches above the top of the fence, I'll string a wire that pulses every three seconds with a hundred volts of electric current. I've been told to smear the wire with peanut butter in order to introduce the deer to the unprecedented and memorable sensation of electric shock, after which they should be gone for good. The electricity will run off a solar panel that sits atop one of the posts, reaching toward the sun like some gigantic high-tech blossom. This last touch strikes me as a nice bit of jujitsu, turning nature's power against a few of her own.

Intervening against the insects is not quite so straightforward, but here too there may be an art that "itself is nature." The key to eliminating an insect from the garden is knowledge: about its habits, preferences, and vulnerabilities. Most chemical pesticides represent a very crude form of knowledge about insects: that, for example, a powerful chemical such as malathion somehow cripples the nervous systems of most organisms, so a little of the stuff should kill bugs but not (presumably) any bigger creatures. Even though this knowledge has been produced by Homo sapiens wearing lab coats, it is not nearly as sophisticated or precise as the knowledge a ladybug, say, possesses on the subject of aphids. The ladybug is not smart, but she knows one thing exceedingly well: how to catch forty or fifty aphids every day without hurting anybody else. If you think of evolution as a three-and-a-half-billion-year-long laboratory experiment, and the gene pool as the store of information accumulated during the course of that experiment, you begin to appreciate that nature has far more extensive knowledge about her operations than we do. The trick is to put her knowledge to our purpose in the garden.

So far, the only way to harness the ladybug gene for aphid capture is by obtaining whole ladybugs, and this can be done through the mail. For about $5 you can order 4,500 ladybugs from a company that specializes in "biological controls." The ladybugs come in a drawstring pouch that can be kept in the refrigerator; spoon out the bugs onto the leaves of infested plants as needed. This particular firm also sells praying mantis egg cases, which can be sewn onto a tree branch near the garden; when the weather warms in spring the nymphs emerge to take up stations on the upper leaves of your plants. Their patience and stillness are extraordinary, as are their reflexes: a praying mantis can snatch most any flying insect right out of the air.

Not all of the biological controls on the market are insects; some are forms of bacteria. One of them—milky spore—supposedly will solve three pest problems at once: grubs, Japanese beetles, and moles. Grubs are the white, wormy-looking larvae of Japanese beetles. They spend the winter and spring underground, where they chew on the roots of grass, leaving dead spots in the lawn. That would be bad enough, but it happens that moles like to dine on these grubs and they ruin lawns tunneling in pursuit of them. The grubs that get away emerge in July as Japanese beetles, scourge of a great many garden plants; the beetles, which were inadvertently introduced into this country several decades ago, can transform a healthy rosebush into a lacy green frame in a matter of days. Milky spore is a bacterial parasite that knows how to infect one insect at one time in its life: the Japanese beetle at the larval stage. The spores, which come in both powdered or granulated forms, should be sprinkled on the lawn in late spring. The grubs will eventually ingest them and die, the moles will go elsewhere in search of grubs, and the Japanese beetles should never appear. According to the catalog, one treatment will last fifteen years.

Biological controls won't solve every pest problem—there are still too few of them, for one thing. But the approach holds promise, and suggests what can be accomplished when we learn to exploit nature's self-knowledge, and stop thinking of our art and technology as being necessarily opposed to nature. For how are we to categorize milky spore disease as a form of human intervention in the landscape? Is it technological, or natural? The categories are no longer much help, at least in the garden.

I won't know for a while whether I've completely solved my pest problem. But, puttering in my newly fenced garden, watching the mantises standing sentry on the tops of my tomatoes and the ladybugs running search-and-destroy missions among the eggplants, I'm starting to feel a lot more relaxed about it. Though Dudleytown remains over the next hill, I know I can stall its advance as long as I continue to put my thought and sweat into this patch of land. I still have much to learn, and there are going to be setbacks, I'm sure; gardening is not a once-and-for-all thing. Yet I think I've drawn a workable border between me and the forest. Might it prove to be a Maginot Line? That's possible, but I think unlikely. Because it doesn't depend on technological invincibility. Nor does it depend on the benignity of nature. It depends on me acting like a sane and civilized human, which is to say, as a creature whose nature it is to remake his surroundings, and whose culture can guide him on questions of aesthetics and ethics. What I'm making here is a middle ground between nature and culture, a place that is at once of nature and unapologetically set against it; what I'm making is a garden.

CHAPTER 3

Why Mow?

No lawn is an island, at least in America. Starting at my front stoop, this scruffy green carpet tumbles down a hill and leaps across a one-lane road into my neighbor's yard. From there it skips over some wooded patches and stone walls before finding its way across a dozen other unfenced properties that lead down into the Housatonic Valley, there to begin its march south toward the metropolitan area. Once below Danbury, the lawn—now purged of weeds and meticulously coiffed—races up and down the suburban lanes, heedless of property lines. It then heads west, crossing the New York border; moving now at a more stately pace, it strolls beneath the maples of Larchmont, unfurls across a dozen golf courses, and wraps itself around the pale blue pools of Scarsdale before pressing on toward the Hudson. New Jersey next is covered, an emerald postage stamp laid down front and back of ten thousand split-levels, before the broadening green river divides in two. One tributary pushes south, striding across the receptive hills of Virginia and Kentucky but refusing to pause until it has colonized the thin, sandy soils of Florida. The other branch dilates and spreads west, easily overtaking the Midwest's vast grid before running up against the inhospitable western states. But neither obdurate soil nor climate will impede the lawn's march to the Pacific: it vaults the Rockies and, abetted by a monumental irrigation network, proceeds to green great stretches of western desert.

Nowhere in the world are lawns as prized as in America. In little more than a century, we've rolled a green mantle of it across the continent, with scant thought to the local conditions or expense. America has some 50,000 square *miles* of lawn under cultivation, on which we spend an estimated $30 billion a year—this according to the Lawn Institute, a Pleasant Hill, Tennessee, outfit devoted to publicizing the benefits of turf to Americans (surely a case of preaching to the converted). Like the interstate highway system, like fast-food chains, like television, the lawn has served to unify the American landscape; it is what makes the suburbs of Cleveland and Tucson, the streets of Eugene and Tampa, look more alike than not. According to Ann Leighton, the late historian of gardens, America has made essentially one important contribution to world garden design: the custom of "uniting the front lawns of however many houses there may be on both sides of a street to present an untroubled aspect of expansive green to the passerby." France has its formal, geometric gardens, England its picturesque parks, and America this unbounded democratic river of manicured lawn along which we array our houses.

To stand in the way of such a powerful current is not easily done. Since we have traditionally eschewed fences and hedges in America, the suburban vista can be marred by the negligence—or dissent—of a single property owner. This is why lawn care is regarded as such an important civic responsibility in the suburbs, and why, as I learned as a child, the majority will not tolerate the laggard or dissident. My father's experience with his neighbors in Farmingdale was not unique. Every few years a controversy erupts in some suburban community over the failure of a homeowner to mow his lawn. Not long ago, a couple that had moved to a $440,000 home in Potomac, Maryland, got behind in their lawn care and promptly found themselves pariahs in their new community. A note from a neighbor, anonymous and scrawled vigilante-style, appeared in their mailbox: *"Please, cut your lawn.* It is a disgrace to the entire neighborhood." That subtle yet unmistakable frontier, where the crew-cut lawn rubs up against the shaggy one, is enough to disturb the peace of an entire neighborhood; it is a scar on the face of suburbia, an intolerable hint of trouble in paradise.

That same scar shows up in *The Great Gatsby,* when Nick Carraway rents the house next to Gatsby's and fails to maintain his lawn according to West Egg standards. The rift between the two lawns so troubles Gatsby that he dispatches his gardener to mow Nick's grass and thereby erase it. The neighbors in Potomac displayed somewhat less savoir faire. Some offered to lend the couple a lawn mower. Others complained to county authorities, until the offenders were hauled into court for violating a local ordinance under which any weed more than twelve inches tall is presumed to be "a menace to public health." Evidently, dubious laws of this kind are on the books in hundreds of American municipalities. In a suburb of Buffalo, New York, there lives a Thoreau scholar who has spent the last several years in court defending his right to grow a wildflower meadow in his front yard. After neighbors took it upon themselves to mow down the offending meadow, he erected a sign that said: "This yard is not an example of sloth. It is a natural yard, growing the way God intended." Citing an ordinance prohibiting "noxious weeds," a local judge ordered the Buffalo man to cut his lawn or face a fine of $50 a day. The Thoreau scholar defied the court order and, when last heard from, his act of suburban civil disobedience had cost him more than $25,000 in fines.

I wasn't prepared to take such a hard line on my own new lawn, at least not right off. So I bought a lawn mower, a Toro, and started mowing. Four hours every Saturday. At first I tried for a kind of Zen approach, clearing my mind of everything but the task at hand, immersing myself in the lawn-mowing here and now. I liked the idea that my weekly sessions with the grass would acquaint me with the minutest details of my yard. I soon knew by heart the precise location of every stump and stone, the tunnel route of each resident mole, the exact address of every anthill. I noticed that where rain collected white clover flourished, that it was on the drier rises that crabgrass thrived. After a few weekends I had in my head a map of the lawn that was as precise and comprehensive as the mental map one has to the back of his hand.

The finished product pleased me too, the fine scent and the sense of order restored that a new-cut lawn exhales. My house abuts woods on two sides, and mowing the lawn is, in both a real and a metaphorical sense, how I keep the forest at bay and preserve my place in this landscape. Much as we've come to distrust it, dominating nature is a deep human urge and lawn mowing answers to it. I thought of the lawn mower as civilization's knife and my lawn as the hospitable plane it carved out of the wilderness. My lawn was a part of nature made fit for human habitation.

So perhaps the allure of the lawn is in the genes. The sociobiologists think so: they've gone so far as to propose a "Savanna Syndrome" to explain our fondness for grass. Encoded in our DNA is a preference for an open grassy landscape resembling the shortgrass savannas of Africa on which we evolved and spent our first few thousand years. A grassy plain dotted with trees provides safety from predators and a suitable environment for grazing animals; this is said to explain why we have remade the wooded landscapes of Europe and North America in the image of East Africa. Thorstein Veblen, too, thought the popularity of lawns might be a throwback to our pastoral roots. "The close-cropped lawn," he wrote in *The Theory of the Leisure Class,* "is beautiful in the eyes of a people whose inherited bent it is to readily find pleasure in contemplating a well-preserved pasture or grazing land."

These theories go some way toward explaining the widespread appeal of grass, but they don't fully account for the American Lawn. They don't, for instance, account for the keen interest Jay Gatsby takes in Nick Carraway's lawn, or the scandal my father's unmowed lawn sparked in Farmingdale. Or the fact that, in America, we have taken down our fences and hedges in order to combine our lawns. And they don't account for the unmistakable odor of virtue that hovers in this country over a scrupulously maintained lawn.

To understand this you need to know something about the history of lawns in America. It turns out that the American lawn is a fairly recent invention, a product of the years following the Civil War, when the country's first suburban communities were laid out. If any individual can be said to have invented the American lawn, it is Frederick Law Olmsted. In 1868, he received a commission to design Riverside, outside of Chicago, one of the first planned suburban communities in America. Olmsted's design stipulated that each house be set back thirty feet from the road, and it prohibited walls. He was reacting against the "high dead walls" of England, which he felt made a row of homes there seem like "a series of private madhouses." In Riverside each owner would maintain one or two trees and a lawn that would flow seamlessly into his neighbors', creating the impression that all lived together in a single park.

Olmsted was part of a generation of American landscape designer/reformers—along with Andrew Jackson Downing, Calvert Vaux, and Frank J. Scott—who set out at mid-century to beautify the American landscape. That it needed beautification may seem surprising to us today, assuming as we do that the history of the landscape is a story of decline, but few at the time thought otherwise. William Cobbett, visiting from England, was struck at the "out-of-door slovenliness" of American homesteads. Each farmer, he wrote, was content with his "shell of boards, while all around him is as barren as the sea beach . . . though there is no English shrub, or flower, which will not grow and flourish here." The land looked like it had been shaped and cleared in a great hurry (as indeed it had): the landscape largely denuded of trees, makeshift fences outlining badly plowed fields, and tree stumps everywhere one looked. As soon as a plot of land was exhausted, farmers would simply clear a new one, leaving the first to languish. As Cobbett and many other nineteenth-century visitors noted, hardly anyone practiced ornamental gardening; the typical yard was "landscaped" in the style southerners would come to call white trash—a few chickens, some busted farm equipment, mud and weeds, an unkempt patch of vegetables.

This might do for farmers, but for the growing number of middle-class city people moving to the "borderland" in the years following the Civil War, something more respectable was called for. In 1870, Frank J. Scott, seeking to make Olmsted's and Downing's design ideas accessible to the middle class, published the first volume ever devoted to "suburban home embellishment": *The Art of Beautifying Suburban Home Grounds,* a book that probably did more than any other to determine the look of the suburban landscape in America. Like so many reformers of that time, Scott was nothing if not sure of himself: "A smooth, closely-shaven surface of grass is by far the most essential element of beauty on the grounds of a suburban house."

Americans like Olmsted and Scott did not invent the lawn—lawns had been popular in England since Tudor times. But in England lawns were usually found only on estates; the Americans democratized them, cutting the vast manorial greenswards into quarter-acre slices everyone could afford (especially after 1830, when Edwin Budding, a carpet manufacturer, patented the first practical lawn mower). Also, the English never considered the lawn an end in itself: it served as a setting for lawn games and as a backdrop for flower beds and trees. Scott subordinated all other elements of the landscape to the lawn; flowers were permissible, but only on the periphery of the grass: "Let your lawn be your home's velvet robe, and your flowers its not too promiscuous decoration."

But Scott's most radical departure from Old World practice was to dwell on the individual's responsibility to his neighbors. "It is unchristian," he declared, "to hedge from the sight of others the beauties of nature which it has been our good fortune to create or secure." One's lawn, Scott held, should contribute to the collective landscape. "The beauty obtained by throwing front grounds open together, is of that excellent quality which enriches all who take part in the exchange, and makes no man poorer." Scott, like Olmsted before him, sought to elevate an unassuming patch of turfgrass into an institution of democracy; those who would dissent from their plans were branded as "selfish," "unneighborly," "unchristian," and "undemocratic."

With our open-faced front lawns we declare our like-mindedness to our neighbors—and our distance from the English, who surround their yards with "inhospitable brick walls, topped with broken bottles" to thwart the envious gaze of the lower orders. The American lawn is an egalitarian conceit, implying that there is no reason to hide behind hedge or fence since we all occupy the same middle class. We are all property owners here, the lawn announces, and that suggests its other purpose: to provide a suitably grand stage for the proud display of one's own house. Noting that our yards were organized "to capture the admiration of the street" one landscape architect in 1921 attributed the popularity of open lawns to "our infantile instinct to cry 'hello!' to the passerby, [and] lift up our possessions to his gaze."

Of course the democratic front yard has its darker, more coercive side, as my family learned in Farmingdale. In commending the "plain style" of an unembellished lawn for American front yards, the mid-century designer/reformers were, like Puritan ministers, laying down rigid conventions governing our relationship to the land, our observance of which would henceforth be taken as an index to our character. And just as the Puritans would not tolerate any individual who sought to establish his or her own back-channel relationship with the divinity, the members of the suburban utopia do not tolerate the homeowner who establishes a relationship with the land that is not mediated by the group's conventions. The parallel is not as farfetched as it might sound, when you recall that nature in America has often been regarded as divine. Think of nature as Spirit, the collective suburban lawn as the Church, and lawn mowing as a kind of sacrament. You begin to see why ornamental gardening would take so long to catch on in America, and why my father might seem an antinomian in the eyes of his neighbors. Like Hester Prynne, he claimed not to need their consecration for his actions; think of his initials in the front lawn as a kind of Emerald Letter.

Perhaps because it is this common land, rather than race or tribe, that makes us all American, we have developed a deep-seated distrust of individualistic approaches to the landscape. The land is too important to our identity as Americans to simply allow everybody to have their own way with it. And having decided that the land should serve as a vehicle of consensus, rather than as an arena for self-expression, the American lawn—collective, national, ritualized, and plain—presented the ideal solution. The lawn has come to express our attitudes toward the land as eloquently as Le Nôtre's confident geometries expressed the humanism of Renaissance France, or Capability Brown's picturesque parks expressed the stirrings of romanticism in England.

After my first season of lawn mowing, the Zen approach began to wear thin. I had by then taken up flower and vegetable gardening, and soon came to resent the four hours that my lawn demanded of me each week. I tired of the endless circuit, pushing the howling mower back and forth across the vast page of my yard, recopying the same green sentence over and over: "I am a conscientious homeowner. I share your middle-class values." Lawn care was gardening aimed at capturing "the admiration of the street," a ritual of consensus I did not have my heart in. I began to entertain idle fantasies of rebellion: Why couldn't I plant a hedge along the road, remove my property from the national stream of greensward, and do something else with it?

The third spring I planted fruit trees in the front lawn, apple, peach, cherry, and plum, hoping these would relieve the monotony and at least begin to make the lawn productive. In back I put in a perennial border. I built three raised beds out of old chestnut barn boards and planted two dozen different vegetable varieties. Hard work though it was, removing the grass from the site of my new beds proved a keen pleasure. First I outlined the beds with string. Then I made an incision in the lawn with the sharp edge of a spade. Starting at one end, I pried the sod from the soil and slowly rolled it up like a carpet. The grass made a tearing sound as I broke its grip on the earth. I felt a little like a pioneer subduing the forest with his ax; I daydreamed of scalping the entire yard. But I didn't do it, didn't have the nerve—I continued to observe front-yard convention, mowing assiduously and locating all my new garden beds in the backyard.

The more serious about gardening I became, the more dubious lawns seemed. The problem for me was not, as it was for my father, with the relation to my neighbors that a lawn implied; it was with the lawn's relationship to nature. For however democratic a lawn may be with respect to one's neighbors, with respect to nature it is authoritarian. Under the Toro's brutal indiscriminate rotor, the landscape is subdued, homogenized, dominated utterly. I became convinced that lawn care had about as much to do with gardening as floor waxing, or road paving. Gardening was a subtle process of give and take with the landscape, a search for some middle ground between culture and nature. A lawn was nature under culture's boot.

Mowing the lawn, I felt like I was battling the earth rather than working it; each week it sent forth a green army and each week I beat it back with my infernal machine. Unlike every other plant in my garden, the grasses were anonymous, massified, deprived of any change or development whatsoever, not to mention any semblance of self-determination. I ruled a totalitarian landscape.

Hot monotonous hours behind the mower gave rise to existential speculations. I spent part of one afternoon trying to decide who, in the absurdist drama of lawn mowing, was Sisyphus. Me? The case could certainly be made. Or was it the grass, pushing up through the soil every week, one layer of cells at a time, only to be cut down and then, perversely, encouraged (with lime, fertilizer, etc.) to start the whole doomed process over again? Another day it occurred to me that time as we know it doesn't exist in the lawn, since grass never dies or is allowed to flower and set seed. Lawns are nature purged of sex or death. No wonder Americans like them so much.

And just where *was* my lawn, anyway? The answer's not as obvious as it seems. Gardening, I had by now come to appreciate, is a painstaking exploration of place; everything that happens in my garden—the thriving and dying of particular plants, the maraudings of various insects and other pests—teaches me to know this patch of land more intimately, its geology and microclimate, the particular ecology of its local weeds and animals and insects. My garden prospers to the extent I grasp these particularities and adapt to them. Lawns work on the opposite principle. They depend for their success on the *overcoming* of local conditions. Like Jefferson superimposing his great grid over the infinitely various topography of the Northwest Territory, we superimpose our lawns on the land. And since the geography and climate of much of this country is poorly suited to turfgrasses (none of which are native), this can't be accomplished without the tools of twentieth-century industrial civilization: its chemical fertilizers, pesticides, herbicides, machinery, and, often, computerized irrigation systems. For we won't settle for the lawn that will grow here; we want the one that grows *there,* that dense springy supergreen and weed-free carpet, that platonic ideal of a lawn featured in the Chemlawn commercials and magazine spreads, the kitschy sitcom yards, the sublime links and pristine diamonds. Our lawns exist less here than there; they drink from the national stream of images, lift our gaze from the real places we live and fix it on unreal places elsewhere. Lawns are a form of television.

Need I point out that such an approach to "nature" is not likely to be environmentally sound? Lately we have begun to recognize that we are poisoning ourselves with our lawns, which receive, on average, more pesticide and herbicide per acre than any crop grown in this country. Suits fly against the national lawn-care companies, and lately interest has been kindled in more "organic" methods of lawn care. But the problem is larger than this. Lawns, I am convinced, are a symptom of, and a metaphor for, our skewed relationship to the land. They teach us that, with the help of petrochemicals and technology, we can bend nature to our will. Lawns stoke our hubris with regard to the land.

What is the alternative? To turn them into gardens. I'm not suggesting that there is no place for lawns *in* these gardens or that gardens by themselves will right our relationship to the land, but the habits of thought they foster can take us some way in that direction. Gardening, as compared to lawn care, tutors us in nature's ways, fostering an ethic of give-and-take with respect to the land. Gardens instruct us in the particularities of place. They lessen our dependence on distant sources of energy, technology, food, and, for that matter, interest. For if lawn mowing feels like copying the same sentence over and over, gardening is like writing out new ones, an infinitely variable process of invention and discovery. Gardens also teach the necessary if un-American lesson that nature and culture can be compromised, that there might be some middle ground between the lawn and the forest—between those who would complete the conquest of the planet in the name of progress, and those who believe it's time we abdicated our rule and left the earth in the care of its more innocent species. The garden suggests there might be a place where we can meet nature halfway.

Probably you will want to know if I have begun to practice what I'm preaching. Well, I have not ripped out my lawn entirely. But each spring larger and larger tracts of it give way to garden. Last year I took a half acre and planted a meadow of black-eyed Susans and ox-eye daisies. In return for a single annual scything, I am rewarded with a field of flowers from May until frost.

The lawn is shrinking, and I've hired a neighborhood kid to mow what's left of it. Any Saturday that Bon Jovi, Judas Priest, or Kiss isn't playing the Hartford Coliseum, this large blond teenage being is apt to show up with a 36-inch John Deere mower that sheers the lawn in less than an hour. It's $30 a week, and I don't particularly like having this kid around—his discourse consists principally of grunts, and he eyes my wife like he's waiting for a *Penthouse* letter to unfold—but he's freed me from my dark musings about the lawn and so given me more time in the garden.

Out in front, along the road where my lawn overlooks my neighbors', and in turn the rest of the country's, I have made my most radical move. I built a split-rail fence and have begun to plant a hedge along it—a rough one made up of forsythia, lilac, bittersweet, and bridal wreath. As soon as this hedge grows tall and thick, my secession from the national lawn will be complete. Anything then is possible. I *could* let it all go to meadow, or even forest, except that I'm not sure I go for that sort of self-effacement. I could put in a pumpkin patch, a lily pond, or maybe an apple orchard. And I could even leave an area of grass. But if I did choose to do that, this would be a very different lawn from the one I have now. For one thing, it would have a frame, which means it could accommodate plants more subtle and various than the screaming marigolds, fierce red salvias, and muscle-bound rhododendrons that people usually throw into the ring against a big unfenced lawn. Walled off from the neighbors, no longer a tributary of the national stream, my lawn would now form a distinct and private place—become part of a garden, rather than a substitute for one. Yes, there might well be a place for a small lawn in my new garden. But I think I'll wait until the hedge fills in before I make my decision.

CHAPTER 4

Compost and Its Moral Imperatives

Soon after we bought this onetime dairy farm and I started reading books about gardening, I began entertaining a fantasy about turning over a shovelful of earth somewhere on the property and finding a thick vein of composted cow manure. To judge from everything I'd read, a trove of this airy, cakelike, jet-black earth would do much more than ensure an impressive harvest; it would elevate me instantly to the rank of *serious* gardener. There isn't an American gardening book published in the last twenty years that doesn't become lyrical on the subject of compost. James Crockett called it "brown gold" in his *Victory Garden,* where he provides a recipe for making compost as complicated as one for a soufflé. The more literary garden writers, such as Eleanor Perényi and Allen Lacy, offer fervent chapters on the benefits and, strange as it sounds, the virtues compost confers. The gardening periodicals—*Organic Gardening* and *National Gardening* in particular—regularly profile heroic gardeners singled out less for the elegant design and lush growth of their perennial borders than for the steaming heaps of compost dotting their yards. In American gardening, the successful compost pile seems almost to have supplanted the perfect hybrid tea rose or the gigantic beefsteak tomato as the outward sign of horticultural grace. What I read about compost gave me my first inkling that gardening, which I had approached as a more or less secular pastime, is actually moral drama of a high order.

Before attempting to grasp the metaphysics of compost, the reader might want briefly to consider the stuff itself. Compost, very simply, is partially decomposed organic matter. Given sufficient time, moisture, and oxygen, any pile of leaves, grass clippings, flower heads, brush, manure, or vegetable scraps will, by the action of bacteria, decay into a few precious shovelfuls of compost. All of the elaborate theories, formulas, and mechanical devices for making compost are really just tricks for speeding this natural process. (A rotating steel drum on the market is said to produce compost in fourteen days; most books say it takes three months.)

Some gardeners, and even some garden writers, talk about compost as if it were fertilizer, but that is only part of the story, and it is somewhat misleading. It is true that compost contains nitrogen, phosphorus, and potash (the principal ingredients of fertilizer), but not in terribly impressive quantities. The real benefits of compost lie in what humus—its main constituent—does to the soil. Consider:

1. Compost improves the soil's "structure." Soil is made up of clay, sand, silt, and organic matter, in varying proportions. Too much clay or silt, and the soil tends to become compacted, making it difficult for air, water, and roots to penetrate. Too much sand, and the soil's ability to retain water and nutrients is compromised. An ideal, friable garden soil consists of airy crumbs in which particles of sand, clay, and silt are held together by humic acid. Compost helps these particles to form.

2. Compost increases the soil's water-holding capacity. One experiment I read about found that 100 pounds of sand will hold 25 pounds of water, 100 pounds of clay will hold 50 pounds, and 100 pounds of humus will hold 190 pounds. A soil rich in compost will need less watering, and the plants growing in it will better withstand drought.

3. Because it is so dark in color, compost absorbs the sun's rays and warms the soil.

4. Compost teems with microorganisms, which break down the organic matter in soil into the basic elements plants need.

5. Because it is made up of decaying vegetable matter, compost

contains nearly every chemical plants need to grow, including such trace elements as boron, manganese, iron, copper, and zinc, not often found in commercial fertilizer. Compost thus returns to the soil a high proportion of the things agriculture takes out of it.

And yet as important as these benefits are, they don't account for the halo of righteousness that has come to hover over compost and those who make it. There are many other sources of humus, after all. To understand compost's mystery, one probably needs to know somewhat less about soil science than about the reasons Americans garden. Which, to judge from the literature and my conversations with experienced gardeners, frequently have less to do with considerations of beauty than of virtue.

Much of the credit for compost's exalted status must go to J. I. Rodale, the founding editor of *Organic Gardening,* who, until his death in 1971, promoted the virtues of organic gardening with a zeal bordering on the messianic. As Eleanor Perényi tells his story in *Green Thoughts,* Rodale was a latter-day Jeremiah, calling on Americans to follow him out of the agricultural wilderness. This is how Perényi, ordinarily the most sober of garden writers, describes her own conversion:

> [Rodale's] bearded countenance glared forth from the editorial page like that of an Old Testament prophet in those days (since his death it has been supplanted by the more benign one of his son), and his message was stamped on every page. Like all great messages, it was simple, and to those of us hearing it for the first time, a blinding revelation. Soil, he told us, isn't a substance to hold up plants in order that they may be fed with artificial fertilizers, and we who treated it as such were violating the cycle of nature. We must give back what we took away.

The way to give back what we had taken, to redeem our relationship with nature, was through compost.

As Rodale himself was the first to admit, there was nothing particularly new about composting. Agriculture had relied on composted organic waste for thousands of years—until the invention, early in this century, of chemical fertilizers. By World War II, most American farmers had been persuaded that all their crops needed in order to thrive were regular, heavy applications of fertilizer. To the farmer, however, the temptations of fertilizer pose something of a Faustian dilemma. At first, yields increase dramatically. But the cost is high, for the chemicals in fertilizer gradually kill off the biological activity in the soil and ruin its structure. Eventually, few organic nutrients remain, leaving crops completely dependent on fertilizer—the soil has become little more than a device to hold plants upright while they gorge themselves on 5-10-5. And to make matters worse, the more fertilizer he uses, the more problems the farmer has with disease and insects, since chemical fertilizer seems to weaken a plant's resistance. After the war, the farmer in this predicament succumbed to a host of new chemical temptations—DDT, Temik, chlordane—and it wasn't long before he found himself deep in agricultural hell.

The home gardener, meanwhile, had been walking down pretty much the same ruinous road, buying more and more chemical fertilizer and then more and more pesticides. By the 1960s, the shelves of his garage were lined with the dubious products of America's petrochemical industry: Cygon, Sevin, kelthane, benomyl, malathion, folpet, diazinon. Where one might reasonably have expected to find the logo of Burpee or Agway there were now the wings of Chevron. Somehow gardening, this most wholesome and elemental of pastimes, had gotten cross-wired with the worst of industrial civilization.

This is the wilderness in which Rodale found the American gardener and confronted him with a stark moral choice: he could continue to use petrochemicals to manufacture flowers and vegetables, or he could follow Rodale, learn how to compost, and redeem the soil—and, the implication was clear, himself.

When Rodale first made his pitch, he was greeted with the degree of respect that is usually accorded prophets. Even as late as the 1960s, he was generally regarded as a crank. When he keeled over and died during a taping of the Dick Cavett show in 1971, the nation responded with a smirk. Johnny Carson told jokes about it for weeks. But as concern over pesticides and the environment deepened during the 1970s, Rodale's message won a wider hearing. Today his is the conventional wisdom in home gardening, and his ideas have even made inroads in American agriculture.

That Rodale should have founded a quasi-religious movement—and that the compost pile should have emerged as a status symbol among American gardeners—makes perfect sense when you consider the attitudes Americans have traditionally held toward the land. The apotheosis of compost is really just the latest act in a long-running morality play about the American people and the American land. In the garden writer's paeans to compost you can still hear echoes of Jefferson's agrarian ideal, paraphrased here by Henry Nash Smith: "Cultivating the earth confers a valid title to it; the ownership of the land, by making the farmer independent, gives him social status and dignity, while constant contact with nature makes him virtuous. . . ."

At least in a metaphorical way, compost restores the gardener's independence—if only from the garden center and the petrochemical industry. With the whole of the natural cycle reproduced in his garden, the gardener no longer has to depend on anyone else (save perhaps the seed merchant) to grow his own food. And because it makes the soil more fertile, composting flatters the old American belief that improving the land strengthens one's claim to it.

This notion of the garden as a realization in miniature of the agrarian ideal seems to have first appeared in the nineteenth century, as Americans in large numbers began leaving the farm for the city. If America could no longer remain primarily a nation of farmers, at least town-dwelling Americans might, by gardening, cultivate some of the rural virtues. "The man who has planted a garden feels that he has done something for the good of the world," wrote Charles Dudley Warner, editor of the *Hartford Courant,* at mid-century. "He belongs to the producers. . . . It is not simply beets and potatoes and corn and string beans that one raises in his well-hoed garden, it is the average of human life." Around the same time, Thoreau planted his bean field at Walden, not so much in order to grow beans that he might eat or sell, but so that he might harvest tropes about the human condition. Improving the soil improved the man.

Americans had come to regard gardening as much more than a pastime, and in the decades prior to the Civil War, horticulture actually attained the status of moral crusade for a time. In an era characterized by "the restlessness and din of the railroad principle," wrote Lydia H. Sigourney in 1840, gardening "instills into the bosom of the man of the world, panting with the gold fever, gentle thoughts, which do good, like a medicine." Addressing the prosperous Bostonians who crowded the Massachusetts Horticultural Society each Saturday morning to hear inspirational talk about gardening and self-improvement, Ezra Weston declared in 1845 that "he who cultivates a garden, and brings to perfection flowers and fruits, cultivates and advances at the same time his own nature."

The hortatory rhetoric may sound foreign to us today, but what about the underlying assumptions? These, it seems to me, we share. No less than the nineteenth-century transcendentalists and reformers, we look to the garden today as a source of moral instruction. They sought a way to preserve the Jeffersonian virtues even in the city; we seek a way to use nature without damaging it. In much the same way that the antebellum garden became a proof of the agrarian ideal, we regard our own plots, hard by the compost pile, as models of ecological responsibility. Under both dispensations, gardening becomes, at least symbolically, an act of redemption.

So pious an attitude toward gardening undoubtedly would strike a

European as absurd. You do not read much about compost in English garden literature. This is partly because the sort of people who write garden books in England are not usually the same sort of people who handle soil. But I think the deeper reason is that British gardeners have traditionally regarded themselves more as artists than as reformers. The issues in English garden writing are invariably framed in aesthetic, rather than moral, terms. Gertrude Jekyll, the influential turn-of-the-century garden designer and writer, borrowed the metaphors of art, not religion, to talk about gardening: she likened plants to a "box of paint" and held that we must "use the plants that they shall form beautiful pictures." *The Education of a Gardener,* by Russell Page, perhaps the most celebrated garden designer of recent times, follows the traditional form of an artist's autobiography, chronicling the artist's discovery of his gift, the development of a personal vision and style, and the various intersections of his life and art. Not a word about compost, self-improvement, or the state of the biosphere.

As might be expected, the gardens made by aesthetes are considerably more pleasing to the eye than those made by moralists. It is no accident that Americans have yet to produce many world-famous gardens or landscape architects, or to found a style of garden design that anyone else would want to copy. I'm not saying we don't have beautiful gardens in this country—we do—but how many of these are derivative of European or Oriental styles? Despite the fact that they seldom work well in our climate or light, we persist in planting copies of English perennial borders—even in the deserts of southern California! So far, at least, American garden design (does the phrase evoke *anything?*) has achieved little of the distinctiveness found in American writing, music, art, or even cooking. Garden design remains the one corner of our culture in which our dependence on England has never been completely broken. Those who care about the look of their gardens still hire English designers (or their imitators) and study English gardening books. Even at this late date, anglophilia continues to rule American gardening.

And yet from the English perspective, some of our most prized gardens scarcely deserve the label. I'm thinking here of Central Park, surely one of the most successful man-made landscapes in America. So how is it that Russell Page can offhandedly dismiss Olmsted's masterpiece as "a stunted travesty of an English eighteenth-century park"? The first time I read this, I bristled at the judgment. But now I think I understand what he means. Even by the relatively informal standards of the English landscape garden on which it is modeled, Central Park is woefully literal and underdesigned (Page faults it for a "total lack of direction"). Yet this radical informality and utter lack of artifice is probably what we like best about it. Central Park pretends not to have been designed. It is less a garden than a counterfeit natural landscape, and New Yorkers seek in it the satisfactions of nature rather than art.

A society that produces "gardens" (or "anti-gardens") like Central Park is one that assumes nature and culture are fundamentally and irreconcilably opposed. And it seems to me that in order to design true gardens of distinction one must have a vision of how the two can be harmonized. It may be this that we lack. Americans have historically tended to regard nature as a cure for culture, or vice versa. Faced with the question of what to do with the land, we always seem to come up with the same crude alternatives: to virtuously subdue it in the name of "progress," or to place it strictly off-limits in "wilderness areas," hallowed places we go seeking an antidote to city life.

A people who believe that nature is somehow sacred—God's second book, according to the Puritans; the symbol of Spirit, according to the transcendentalists—will probably never feel easy bending it to their will, and certainly not for aesthetic reasons. Indeed at least since the time of Thoreau, Americans have seemed more interested in the idea of bending *themselves* to nature's will, which might explain why this country has produced so many more great naturalists than great gardeners. We evidently feel more comfortable taking moral instruction in bean fields and at the feet of trees than arranging plants into pleasing compositions.

We even seem to approach our gardens as naturalists. Consider the typical American gardening book, which is organized like a field guide, plant by plant. It is much less often that you find rock gardens, herbaceous borders, or annual beds considered whole, as you do in English garden books. Instead, each cultivar is given its due, considered as an individual, its habits, character, and flaws appraised. "Flowers one can like or even love for themselves," wrote Katherine White, for many years the *New Yorker*'s garden columnist, "but gardens inevitably relate to man." Alas. It is as if making gardens were somehow unfair to the plants in them, a denial of their individuality and freedom. How long can it be before Americans rally behind the banner of plants' rights?

But back to compost. Eventually I did find the buried treasure. I was digging around the barn one day last fall when suddenly my spade slipped through a patch of particularly airy soil. I turned over a chunk of sod, and there it was: the blackest earth I'd ever seen. I was elated, but only momentarily. Because by then I had read enough about compost to know that finding it didn't really count. Yes, it would be a boon to my vegetables and perennials. But this was a one-time windfall, the moral equivalent of finding a deposit of fossil fuel. I didn't even mention the strike to any of my serious gardening friends. For I now understood that if I wanted to perfect my gardening faith I would have to begin my own compost pile.

Which I promptly did. I built a slatted box out of some scrap lumber, found a shady spot for it (so the compost wouldn't dry out in the sun), and after the first frost had finished off the warm-weather plants, I piled the box high with blackened bean vines, squash leaves, zinnias, sunflower stalks, corncobs, and half a dozen club-sized zucchinis that had eluded timely harvest. I topped off the pile with a shovelful of the compost I'd found (in order to introduce the necessary microorganisms, it's best to begin a compost pile with a bit of compost, on the same principle as sourdough bread making). I mixed it all up, hosed it down, and forgot about it.

By the time I returned to the compost pile in April, I had read enough about American gardening to know that composting was a pretty silly fetish. It would never produce a beautiful perennial border, just a morally correct one, and wasn't that a little absurd? Well, I guess it is, but when I lifted off the undecayed layer of leaves on top and ran my hand through the crumbly, black, unexpectedly warm and sweet-smelling compost below, I felt like I'd accomplished something great. If fertility has a perfume, this surely was it. Mixed in were incompletely composted bits and pieces—vague brown shards that I could still make out as former corncobs and sunflower seed heads. They looked like the shadows of last year's harvest. I have to admit, I was starting to see tropes. This heap of rotting vegetable matter looked more lovely to me than the tallest spike of the bluest delphinium. Right then I realized that, like it or not, I was an American gardener, likely to cultivate in the garden more virtue than beauty.

Summer

CHAPTER 5

Into the Rose Garden

Preparing a bed for roses is a little like getting the house ready for the arrival of a difficult old lady, some biddy with aristocratic pretensions and persnickety tastes. Her stay is bound to be an ordeal, and you want to give her as little cause for complaint as possible. All of a sudden the soil that's served you well for years seems lacking, its drainage dubious and pH off. So I've been double-digging, hauling bales of peat moss, and blowing all at once and in one place the precious cache of compost it's taken me years to accumulate. Up to now, I've avoided growing roses (*real* roses, that is—I've always had a tough climber or two); after all, who courts such captious and intimidating guests? But this spring for some reason the ripe catalog shots of roses I always used to sail right past took hold in my imagination, and I decided to take the plunge.

I think it must have been the two-page spread of "old-fashioned roses" in the Wayside catalog that first seduced me. Here were a dozen ladies (and one debonair gent: Jacques Cartier) that looked nothing like roses were supposed to look. Instead of the trite, chaste, florist-shop bud, these large, shrubby plants bore luxuriant blooms that seemed to cascade down from the page: unruly masses of flower petals—*hundreds* of petals in some of them—just barely contained by form, which in most cases was that of a rosette or a teacup's half globe. Wayside likes to photograph their rose blossoms when they are well along, almost over the top, and then they crop the pictures to make it seem as though the blooms are breaking out of their frames, pushing forward, almost in your face. The whole effect is vaguely lascivious.

But then the names of these roses push you in another direction, back toward the drawing room. Meet Madame Hardy. Please make the acquaintance of Madame Isaac Perrier. *Je voudrais présenter* La Reine Victoria, Belle de Crécy, the Königin von Danemark. Names that withhold more than they disclose, *society* names, their full significance revealed only to those in the know. You do remember Madame Hardy, the widow of Monsieur Hardy? Didn't he used to tend Empress Josephine's rose garden at Malmaison? Yes, of course. The distinguished rosarian Graham Stuart Thomas draws you aside to confide that Madame Hardy is from the damask family, though some of her relatives are centifolia. "There is just a suspicion of flesh pink in [her] half-open buds," he whispers; don't you find her "sumptuous and ravishing"? Hard to tell, here among this group of old roses, whether you've stumbled into a Second Empire drawing room or a Left Bank brothel—whether it is their pedigree or sexuality that gives these ladies their allure.

Dazzled, smitten, I ordered four old roses from Wayside. Madame Hardy, of course; Jacques Cartier, an 1868 introduction who looks quite suave in all his pictures; Königin von Danemark ("a jewel beyond price" and so presumably a bargain at $17.75); and Blanc Double de Coubert, an 1892 hybrid rugosa that Gertrude Jekyll says is "the whitest rose known." From another firm, Roses of Yesterday and Today in California, I ordered a rose Wayside doesn't carry: Maiden's Blush, a shell-pink alba about which the catalog copy is unequivocal: "Nature has created nothing more exquisite in plant or bloom." Maiden's Blush is the oldest rose I ordered; it first bloomed in the fifteenth century. And then I ordered a single modern rose, Queen Elizabeth, a clear pink hybrid bush introduced when Elizabeth was crowned in 1953. With the exception of the prominent English rose breeder David Austin, who finds her somewhat "coarse," most authorities agree that Queen Elizabeth is one of the best roses of this century.

While I awaited the arrival of this sextet by UPS, I prepared their bed and read up on roses. I was soon reminded of the reasons for my former reluctance to grow roses, as book after book retold the familiar horror story of their multitudinous afflictions. The catalog of things that could go wrong was daunting. I had better plan to dress my roses warmly for the winter, I read, for without protection all but the hardiest would succumb to the freezings and thawings of a Connecticut January. And in summer, my roses would insist on plenty of water (an inch a week straight through the hot months), yet because they "don't like to get their feet wet" the drainage of their soil had better be impeccable. So I dug the bed to a depth of two feet and added large quantities of organic matter: compost, aged cow manure, and peat moss. But even a perfect bed will not necessarily halt the battalions of pests and diseases who have singled out the rose for conquest.

The catalog of these occupies fully eight depressing pages in *America's Garden Book,* the New York Botanical Garden's reference guide. Here was a whole host of new worries I was about to invite into my life. I would know a rose had contracted black spot when the leaves developed said black spots, then turned yellow and dropped off. This can't be cured, according to the book, but prevention (with regular applications of fungicide) might be possible. I'm afraid they can't say as much for bronzing, a baffling disease for which nothing can be done. And then there's brown canker, stem canker, leaf rust, powdery mildew, crown gall, and a large assortment of nasty viruses.

If a rose escapes these dismal fates, others lie in wait: rose aphids want to suck the bodily fluids from my plant, if not to the point of death, then certainly long enough to rob its beauty and increase its vulnerability to all the diseases listed above. Later in the season descend hordes of sleek green Japanese beetles, which can swiftly reduce a healthy rose bush to a rickety skeleton. Should the beetles leave anything behind, prepare for the arrival of the leaf rollers, red spider mites, rose chafers, rose curculios, rose midges, rose scales, rose slugs, rose stem borers, and rose weevils. What other plant has had so many insects named in its honor? Enough to make one wonder if the true raison d'être of the rose, its real ecological purpose, is to make a meal for its legion of eponymous pests.

More likely, though, most of these plagues and insect armies didn't enter on the scene until the development of the modern hybrid rose. In recent years hybrid teas have been bred so singlemindedly for looks that resistance to disease and general sturdiness of constitution have been neglected to the point where these plants are virtual basket cases. They're like the inbred offspring of an old royal family: soft, susceptible, too feeble to make their own way in the world. Faint hothouse characters, incapacitated sovereigns for whom we gardeners, fortified with the full panoply of Ortho products, must step in and act as regents.

Not a job I wanted, which is probably the main reason I've stayed away from roses so long. Most nurseries offer nothing but hybrid teas, and these just didn't fit my view of what gardening was supposed to be about. In the balance between nature and civilization that gardening aims to strike, the rose (or at least the gaily packaged Jackson & Perkins roses on sale at my garden center) lists too far toward the side of civilization. Modern roses simply can't get by without the crutch of the chemical industry. I remember my grandfather's roses, wreathed in their white clouds of Ortho Rose Dust, and giving off a perfume distinctly more chemical than floral. (Though Rose Dust, like chlorine, was always a pleasantly evocative scent of summer, at least until I discovered it was poison.) The modern roses he grew, and the ones he gave my parents, never did much for me: perhaps because their blooms were such familiar visual clichés, they scarcely registered. The classic, tight vermilion buds of a Mr. Lincoln made for a hackneyed passage in the garden, a platitude I didn't even hear.

Not that the modern rose lacks for novelty—indeed, novelty is a big part of their problem. Twentieth-century capitalism discovered the rose and decided what it needed after several millennia of successful cultivation was a full-tilt program of R&D, innovation, market research, positioning, and advertising. As gardeners are fond of pointing out, the modern rose industry appears to have modeled itself after Detroit. Each year it introduces a handful of "exciting" new models, many of them in improbable neon and metallic shades better suited to a four-door than a flower, and each bearing a loud, hypey name dreamed up on Madison Avenue and duly trademarked. Chrysler Imperial is actually the name of a rose. So is Sunsation. And Broadway (a two-toned wonder gaudy as a showgirl). Hoola Hoop. Patsy Cline. Penthouse. Sweetie Pie. Twinkie. Teeny Bopper. Fergie. Innovation Minijet. Hotline. Ain't Misbehavin'. Sexy Rexy. Givenchy. Graceland. Good Morning America. And Dolly Parton (a rose with, you have probably guessed, exceptionally large blossoms). It seems to me that the world conjured by these roses is precisely the one we come to gardening to escape.

This was roses, or so at least I thought until I wandered into the quirky, rarefied realm of the "old-fashioned" roses. Here I discovered roses that looked nothing like those of my childhood—or the ones on sale at my garden center—and that seemed to suffer from few of the hybrid's flaws. In the unexpectedly cantankerous world of old roses, I found that I was scarcely alone in my dislike for hybrid teas—in fact such disdain is considered a mark of both common sense and refined taste. Here I ran into eminent American garden writers, like Eleanor Perényi, who speak of their disgust with the "planned obsolescence" of modern roses and go on about the "unforgettable perfume of old roses." But it's the English who've really got the critique of hybrids down: modern roses, as Vita Sackville-West used to complain to anyone who would listen, are insufficiently subtle, too highly colored, altogether too . . . bourgeois. Sackville-West was the forerunner of a whole faction of rose Tories who agitate (decorously, of course) against hybrid roses and cling to dreams of a restoration for their beloved albas, gallicas, damasks, bourbons, and centifolias. The world of roses, which I had always thought of as a mild province inhabited by harmless old biddies and benign rosarians, turns out to be a site of seething conflict.

Today the leading spokesman for the old rose faction is the English rosarian Graham Stuart Thomas, author of *The Old Shrub Roses,* a classic polemic masquerading as genteel garden chat. Thomas has been passionate on the subject of old roses since the 1940s when, as a researcher for the Cambridge Botanic Garden, he began collecting and preserving specimens of old roses then on the verge of extinction. Thomas, and acolytes such as David Austin, make several arguments on behalf of the old rose. Old roses are undeniably hardier than modern roses, and most of them are not nearly as susceptible to the various rose plagues. Old roses also possess a much more powerful perfume—scent, like disease resistance, having been more or less ignored by the modern breeder in his quest for novel colors and perpetual bloom. But here is where the modern rose enjoys an edge. As Thomas acknowledges, old roses offer only a narrow range of colors (from white to pink; no vermilion or yellow to speak of). And, as no garden catalog will fail to remind you (the hybrid teas don't need learned advocates; they have the entire nursery industry and most of the population on their side), the hybrids, unlike most old roses, bloom all season long.

Proponents of the old rose have more than disease resistance and a nice smell on their side, however. Their champions may not acknowledge it directly, but a large part of the appeal of old roses—which seem to be enjoying a renaissance today—is based on snobbery. The war of the roses is at bottom a class war.

The tracts of old-rosarians bristle with the fine distinctions, winks, and code words by which aristocrats have always recognized one another. Vita Sackville-West, in her introduction to Thomas's book, writes that the old rose is "a far quieter and more subtle thing, but oh let me say how rewarding a taste it is when once acquired." She goes on to compare old roses to oysters. Thomas himself damns the hybrid tea with faint praise, acknowledging that they are capable of "producing very finished florist's flowers." David Austin observes that the hybrids are "gross feeders" and that the vermilion of some recent introductions, while admittedly beautiful, is "perhaps a little foreign to the rose."

It is when rosarians are chronicling the history of the rose that issues of class, never far from the gardening world, bubble up to the surface. In their hands, the history of the rose is a thinly disguised parable of class struggle in Europe, as told from the perspective of a superannuated aristocracy. Reading their accounts you begin to sense that, for a plant, the rose carries an awful lot of cultural and political baggage.

Rose history is an extraordinarily complicated subject, and I will not bore you with a lot of it. Suffice it to say that, prior to 1789, the rose world in the West was dominated by a small handful of "families" that had enjoyed unchallenged supremacy for centuries—not unlike the ruling nobility of Europe. The principal rose families included: the gallicas (the preeminent rose during the Roman empire); the damask (a medieval cross between the gallica and a wild rose); the albas (a damask crossed with another species rose, *Rosa canina* or the dog rose); the centifolias, also known as cabbage roses (the ones favored by Dutch Renaissance painters) and the moss roses, which are thought to be an offshoot of the centifolia family. From the Roman Empire through the Enlightenment, members of these five royal families more or less ran the show in Europe. Change was seldom, though not unheard of: now and then two great families would come together in marriage and thereby found a new line, as when damask and alba joined to launch the centifolia in seventeenth-century Holland.

This was the rose world's ancien régime, and it was to last no longer than France's own. For in 1789, as Graham Thomas ruefully notes, "the rose was to suffer a great revolution in common with its most ardent admirers of that time." The upheaval in the rose world was caused by the introduction in Europe of the China rose, *R. chinensis,* a rose with the ability to flower more than once a season. The fact that the ancien régime roses flowered but once had never been considered a flaw until the discovery of the China rose which, though not winter hardy in Europe, could flower all summer long. The rose world suddenly faced a crisis of rising expectations, and it wasn't long before the old families gave way to a generation of new roses that could flower repeatedly. The first of these remontant (recurring) roses was the Portland (of which

Jacques Cartier is one). But the most important rose of this class, the bourbon, did not appear in France until 1823. On the Île de Bourbon, a small island in the Indian Ocean near Mauritius, farmers used to plant a mixed hedge of Autumn Damask and Old Blush China roses. Jean-Baptiste Bréon, a plantsman visiting from Paris, discovered a naturally occurring hybrid of the two varieties, known to islanders as the Edward Rose, and he sent it back to Paris, where it soon became the most popular rose of its time.

The introduction of the China rose to Europe "provided great opportunities," as David Austin judiciously observes in *The Heritage of the Rose,* "but, as is so often the case with such opportunities, also certain dangers." Here was a precarious moment, truly parlous times for the rose, but fortunately (at least from Austin's point of view), the new roses proved to be exceptionally well behaved: he judges the bourbon rose as "the best of both worlds," having brought the remontant trait to Europe without destroying the beauty of the Old World blooms. These new roses may have toppled the ancien régime, but thankfully their own demeanor turned out to be decidedly more aristocratic than Jacobin. A new nobility was installed, a group that had won rather than inherited its position. Here was new blood, true, but it displayed imperial rather than democratic aspirations. The rose's Napoleonic era had begun.

The label is all the more appropriate for the fact that Empress Josephine played an important role in promoting the rose's nineteenth-century golden age. "There was such a gathering together of roses at La Malmaison," Thomas writes, "as had never before been seen." All the old roses were represented, and many new ones—including the Portland and the Noisette—were collected and developed here under the guidance of André Dupont, Josephine's head gardener. Her rose garden was so famous and well regarded that when British warships seized French vessels carrying plant materials (Napoleon's troops were under instructions to scour the world for new rose specimens) they would allow them to pass. Josephine's passion for the rose helped make it the preeminent ornamental flower of the nineteenth century; before now, its fame rested as much on its fragrance and medicinal properties as its beauty. It was also Josephine who set Pierre Joseph Redouté to work on his monumental series of rose paintings, many of which were done at Malmaison. Thus the year 1789 was doubly significant for the rose: it marked the introduction of the China rose in Europe and, with the French Revolution, set off a chain of events culminating in the rose's shining moment at Malmaison. From here on, according to our authorities, it's a long downhill slide to Dolly Parton.

With the rise of the middle class to power in France and England the rose began its decline, gradually coming to assume its familiar modern appearance. The same inventive and competitive impulses that helped spur the industrial revolution were brought to bear on the rose, and breeders for the first time deliberately sought to develop new hybrids with specific traits designed to appeal to the marketplace. And for the first time in the history of gardening, that marketplace was dominated by the middle class, whose particular tastes and requirements soon led to revolutionary changes in the rose—to the point where, in David Austin's words, it became "to all intents and purposes, a new flower."

This new flower had several distinguishing characteristics. First, beginning with the hybrids introduced at mid-century, the rose is no longer a shrub, but a bush. Middle-class gardeners simply did not have the space an old rose commands (many of them grow to a height of six feet); the new market called for blooms, not big shrubs, and the hybridizers hurried to oblige. With the development of the hybrid perpetual, whose lineage rosarians despair of sorting out (one snorts that they are merely "an amalgamation of various roses with certain objectives in view"), the shape of the rose plant—now "clumsy," "ungainly," "too upright"—is completely sacrificed to the period's consuming obsession with the bloom.

This obsession was no doubt stoked by the Victorian fad for flower shows, to which thousands of amateur rose fanciers rushed to enter their prize specimens in competition. Since only blooms could be submitted for judging, it wasn't long before rose breeders stopped paying attention to everything else, a development naturally deplored by the Faction. "Such roses were no doubt very fine when seen on the show bench," huffs one rosarian, "but as garden plants they left much to be desired." But don't for a moment believe this fellow *really* considers any of these blooms "very fine," either. The Victorian passion for novelty had grabbed hold of the rose, and breeders brought forth a range of gaudy new colors (and a few stripes) that lovers of old roses rail about to this day.

In 1867, the very same year that the Second Reform Bill extended the suffrage to the middle classes in England, a French breeder by the name of Guillot crossed a hybrid perpetual named Madame Victor Verdier with a tea rose (the tender crosses of bourbons and China roses) named Madame Bravy to produce La France, generally recognized as the first hybrid tea. This new rose proved to be just what the middle class had been waiting for. It was a petite bush (rarely more than three feet tall), which suited it to small gardens, and it bloomed like crazy—with "no thought to the future," in the alarmed words of one Faction member. Also, with its long, shapely bud, the hybrid tea was destined to triumph on the show bench, since it was at the bud stage that roses were usually exhibited. "The pointed bud of the hybrid tea . . . can be of exquisite beauty," David Austin observes. But there is a high price to pay: "Unfortunately, the open flower tends to be shapeless and lacking in quality—a jumble of petals of no particular form." The rose, in form, is no longer a rosette but a bud just beginning to open. It is at this point that the image that the word "rose" summoned in the mind of Shakespeare's audience gives way to the very different picture the word evokes for us.

The size and constant bloom of the hybrid tea meant that they could be enlisted in Victorian bedding schemes—the dubious practice of planting masses of a single brightly colored flower in variously shaped beds that were then combined to form complex designs. Persian carpets were popular; so were coats of arms; the less ambitious planted half-moons and even tadpoles. The mainstays of these bedding schemes, of course, were annuals, which now could be imported from the tropics and grown in quantity in the new heated glass greenhouses. These flashy, free-blooming upstarts no doubt exerted a powerful competitive effect on rose breeders, who pressed the rose to give more and more bloom in ever brighter colors.

By now the rose plant was little more than a tripod to hold these fat gaudy blooms aloft. The once-noble shrub had been dwarfed, dubbed with a series of undignified names, dressed up in ridiculous hues, and made to stand shoulder to shoulder in a crowd, the glory of its individual blooms subordinated to a mass effect—to a crude "blaze of color." As for this elevation of the bloom above the shrub and its traditions—well, isn't that just like the parvenu, to flaunt the trappings of wealth while paying no heed to its substance? To *breeding*? Yes, yes, the modern rose certainly has *pretensions* to nobility—just look at those names: Princess Grace. John F. Kennedy. Cary Grant. Noble, perhaps, but so . . . so *nouveau.* Arrivistes, all of them! (Jack Kennedy, the rosarian sputters, you're no Jacques Cartier.) And what about Dolly Parton? Barbara Mandrell? *Graceland?* Horrifying, indeed. Is it any wonder that the old rose man went underground for nearly a century, retreating to his country estate, there to pen his acid tracts and despair of the age?

The apes had ransacked the temple of gardening and ravished fair rose.

As I said: an awful lot of baggage for a flower to carry. I thought about this the afternoon my old roses arrived by UPS truck, because they seemed far too frail to bear so much significance. Little more than sticks, they came bare-rooted and swaddled in last week's *Pennysaver,* looking more like habitués of Ellis Island than of Malmaison. The roses were completely dormant; apart from a faint swelling at the buds, they looked dead. Hard to believe I had dropped a total of seventy-five bucks on these twigs, let alone that Western Civilization had had so much to say about them.

The instructions said to soak them overnight in a bucket of warm water, and plant them as soon as possible. The next morning, I dug holes eighteen inches deep and across, and four feet apart. To each hole I added several handfuls of compost, well-rotted cow manure, and peat moss, mixing these together by hand. Then I molded this black, cakey mixture into a mound at the bottom of each hole to serve as a kind of pillow for my fragile charges. The plants look like two octopi joined at the head, the roots coming out one side and the canes the other. The idea, according to the directions, is to set the head (really the bud union) on top of this soil cushion and spread the roots down along the sides.

Here in zone five, Wayside recommends burying the bud union two inches beneath the surface to protect it from winter stresses. I checked the depth, filled the holes with soil, and soaked them deeply to ensure complete contact between the roots and the surrounding earth. Then I tamped the ground around the plants with my foot. After a few days of moisture, the roses would break their dormancy. The roots would send their delicate tentacles deep into the underlying mound and the alchemy by which the rose promised to translate this black mass of manure and decayed vegetable matter into blooms of legendary beauty would begin.

That seemed pretty far off, though, as I stepped back to examine my handiwork. It didn't look like much: six clutches of rose cane jutting at skewed angles out of the mud. Not very sexy, hardly upscale, and not the least bit evocative. No, here was the rose shorn of all its associations, its burden of metaphor—here, it seemed, one could glimpse the rose with what Wallace Stevens called "a mind of winter," as it *was,* without tropes. Without Shakespeare. Without the War of the Roses . . . the crown of thorns . . . *rosy-fingered dawn* . . . sub rosa . . . *Rose is a rose is a rose* . . . the rosary . . . the rosicrucians . . . *The Romance of the Rose* . . . the Rose Bowl . . . the bed of roses . . . *by any other name would smell as sweet* . . . Dante's yellow rose of Paradise . . . *when the fire and the rose are one* . . . the run for the roses . . . *toward the door we never opened/into the rose garden* . . . through rose-colored glasses . . . Rosebud . . . Tennyson's *white rose of virginity* . . . Aphrodite's flower . . . the Virgin

Mary's, too . . . blood of Adonis . . . symbol of love . . . purity . . . transience . . . eternity . . . symbol, it almost seems, of symbols.

All this the forlorn, stubby skeletons seemed to hollow out, render faintly ludicrous. Here it was: a plant, a briar. Period.

After only a few days the buds reddened and swelled, and by the end of two weeks the canes had unfurled around themselves a deep green cloak of leaves, paler, daintier, and in finish more matte than the high-gloss foliage of modern roses. I had read that most old roses flower on "old wood" (last season's growth), so I had no expectation of blooms that first season. But in late June, after a month of rapid growth, Madame Hardy sent forth a generous spray of buds.

I had by now read so much about old roses that I frankly doubted they could live up to their billing. But Madame Hardy was beautiful. From a small, undistinguished bud emerged a tightly wound bundle of pure, porcelain-white petals that were held in a perfect half-globe as if by an invisible teacup. The petals were innumerable yet not merely a mass; more ladylike than that, the fine tissue of Madame Hardy's petals was subtly composed into the quartered form of a rosette, and the blooms made me think of the rose windows of Gothic cathedrals, which had not before looked to me anything like a rose.

It was hard to look at Madame Hardy plain, hard not to think of her as an expression of another time—which of course, as much as being an expression of nature, she is. Though Madame Hardy did not appear until 1832 (bred, you'll recall, by Josephine's rose gardener and named for his wife), she embodies the classic form of old roses, and comes closer to the image the word rose has conjured in people's minds for most of Western history than does the rose in our florist shops today. When Shakespeare compared his love to a rose, this must have been pretty much what he had in mind. To look closely at the bloom of an antique rose is, at least in some small way, an exercise of the historical imagination: you see it through your own eyes, yet also through the eyes of another time. What an odd thing, though, for a rose is not a poem, or a painting, but a flower, part of nature: timeless. Yet man in some sense made Madame Hardy, crossed and recrossed it until it reflected his ideal of beauty—and so today in my garden it reflects the sensibility of another time back at me. The rose is part of nature, but also part of us. So much for the mind of winter.

Admiring the beauty of Madame Hardy, I began to see why she should so excite rosarians of a snobbish bent—and to accept the slightly uncomfortable fact that, at least in the war of the roses, my own sympathies were not with the party of the people. Compared to modern roses, Madame Hardy is indeed an aristocrat, incomparably more subtle and, in form, so much more *poised.*

Once you have grown roses, you can begin to understand why people might project metaphors of social class onto them. Each bush itself forms a kind of social hierarchy. Beneath Madame Hardy's bud union is the rootstock of another, tougher variety—not a hybrid but a crude species rose, some hardy peasant stock that can withstand bad winters, but whose meager flowers interest no one. The prized hybrid is grafted onto the back of this anonymous rootstock, which performs all the hard labor for the rose, working the soil, getting its roots dirty so that the plant may bloom. The prickly shrub itself is not distinguished particularly, but it too is necessary to support the luxury of the bloom—its great mass of leaves manufactures the food, and its branches form the architecture without which flowering would not be possible. And the extravagant, splendid blooms, like true aristocrats, never seem to acknowledge the plant that supports them, or the fact that their own petals were once mere leaves. They comport themselves as though their beauty and station were god-given, transcendent. You cannot discern in the bloom of a rose the work of the plant, the sacrifice of its chafer-eaten leaves, the stink of the manure in which it is rooted. Roots? Madame Hardy asks, ingenuously. What roots?

But if Madame Hardy calls attention to her pedigree, Maiden's Blush, the alba I planted beside her in my garden, seems to press her sexuality on us. Her petals are more loosely arrayed than Madame Hardy's; less done up, almost unbuttoned. Her petals are larger, too, and they flush with the palest flesh pink toward the center, which itself is elusive, concealed in the multiplication of her labial folds. The blush of this maiden is not in her face only. Could I be imagining things? Well, consider some of the other names by which this rose is known: Virginale, Incarnata, La Séduisante, and Cuisse de Nymphe. This last is what the rose is called in France where, as Vita Sackville-West tells us, blooms that blush a particularly deep pink are given the "highly expressive name" of Cuisse de Nymphe Émue, which she demurs from translating. But there it is: the thigh of an aroused nymph.

No, Maiden's Blush is certainly not the kind of old lady I expected when I planted roses. Her concupiscence, in fact, has made me wonder if all the baggage with which the rose has been loaded down might be just a cover for these nymph thighs, for this unmistakable carnality. For though Maiden's Blush bears an especially provocative bloom, every one of the old roses I planted, and all the ones I've since seen and smelled, have been deeply sensuous in a way I wasn't prepared for. Compared to the chaste buds and modest scent of the modern roses, these old ones give freely of themselves. They flower all at once, in a single, climactic week. Their blooms look best fully opened, when their form is most intricate; explicit, yet still so deeply enfolded on themselves as to imply a certain inward mystery. And their various perfumes—ripe peaches, burnt almonds, young chardonnays, even musk—can be overpowering. More even than most floral scents, the fragrance of these roses is impossible to get hold of or describe—it seems to short-circuit conscious thought, to travel in a straight line from nostril to brain stem. Inhale deeply the perfume of a bourbon rose and then try to separate out what is scent, what is memory, what is emotion; you cannot pull apart the threads that form this . . . this *what?*

By the time all my old roses had bloomed I had begun to think that maybe Marx has less to tell us about the world of roses than Freud. Certainly Freud would assume that anything we have invested with this much significance must exert some powerful sexual pull. I returned to my rose literature, and surely enough, the same rosarians whose prose had seemed to bristle with class consciousness now read to me as slightly sex-crazed. Would it be disrespectful to suggest that Graham Stuart Thomas, O.B.E., V.M.H., D.H.M., V.M.M., has a thing for old roses? Here is his full description of Madame Hardy: "There is just a suspicion of flesh pink in the half-open buds, emerging from their long calyces, and the flower open-cupped, rapidly becoming flat, the outer petals reflexing in a most beautiful manner, leaving the center almost concave, of pure white, with a small green eye . . . sumptuous and ravishing." The scent of Maiden's Blush reduces Sir Thomas to the rapturous ineffables of a trashy romance writer: her blooms are "intense, intoxicating, delicious . . . my senses have not yet found the means of conveying to my pen their qualities." Marie Louise, a rose raised at Malmaison in 1813, brings out the Humbert Humbert in him: "To lift up the leafy sprays and look steadily at the fully opened blooms is a revelation. . . ." I was beginning to understand why rosarians tend to be men. Men, and then of course Vita Sackville-West, who could certainly work herself up writing about old roses: "Rich they were, rich as a fig broken open, soft as a ripened peach, freckled as an apricot, coral as pomegranate, bloomy as a bunch of grapes . . ." Your opinion, Doctor Freud?

If the allure of old roses is in the frank sensuality of their blooms, then what are we to make of the development and eventual triumph of the modern hybrid tea? Maybe the Victorian middle class simply couldn't deal with the rose's sexuality. Perhaps what really happened in 1867 was a monumental act of horticultural repression. By transforming the ideal of rose beauty from the fully opened bloom to the bud, the Victorians took a womanly flower and turned her into a virgin—a venerated beauty when poised on the verge of opening, but quickly fallen after that.

As for the prized new trait of continual bloom, that too can be seen as a form of sublimation. For the hybrid roses don't give more bloom, really, they just parcel their blooms out over a longer period; they save and reinvest. So instead of abandoning herself to one great climax of bloom, the rose now doles out her blossoms one by one, always holding back, forever on the verge, never quite . . . finishing. The idea of a flower that never finishes would have struck the Elizabethans as perverse; one of the things they loved most about the rose was the way it held nothing back, the way it bloomed unreservedly and then was spent. But the Victorians bred this sexual rhythm out of the rose, subordinating it to the period's cult of virginity, as well as its new concepts of economy. From them we've inherited a girlish flower, pretty perhaps, but scrubbed to the point of scentlessness, no more alluring or sexually aware than a girl scout.

To look at a flower and think of sex—what exactly can this mean? Emerson wrote that "nature always wears the colors of the spirit," by which he meant that we don't see nature plain, only through a screen of human tropes. So in our eyes spring becomes youth, trees truths, and even the humble ant becomes a big-hearted soldier. And certainly when we look at roses and see aristocrats, old ladies and girl scouts, or symbols of love and purity, we are projecting human categories onto them, saddling them with the burden of our metaphors.

But is there any other way to look at nature? Thoreau thought there was. He plumbed Walden Pond in winter in order to relieve nature of precisely this human burden—to "recover the long lost bottom of Walden Pond" from the local legends that held it bottomless. Thoreau was confident he could distinguish between nature (the pond, which he determined had a depth of exactly one hundred and two feet) and culture (the stories people told about its bottomlessness); he strove to drive a wedge between the two once and for all—to see the pond with a mind of winter: unencumbered, as it really was. "Let us settle ourselves, and work and wedge our feet downward through the mud and slush of opinion, and prejudice, and tradition, and delusion, and appearance, that alluvion which covers the globe . . . till we come to a hard bottom and rocks in place, which we can call *reality,* and say, This is . . ." The transcendentalists looked to nature as a cure for culture, but before it can exert its "sanative influence," we have first to scrape off the crust of culture that has formed over it.

This neat segregation of nature and culture gets complicated when you get to garden plants such as the rose, which perhaps begins to explain why Thoreau preferred swamps to gardens. For the rose not only wears the colors of our spirit, it *contains* them. Roses have been "cultivated" for so long, crossed and recrossed to reflect our ideals, that it is by now impossible to separate their nature from our culture. It is more than a conceit to suggest that Madame Hardy's elegance embodies something of the society that produced her, or that Graceland's slickness embodies something of ours. To a certain extent, the same holds true for all hybrid plants, but no other has received as much sustained attention from the hybridizer, that practitioner, in Shakespeare's words, of an "art [that] itself is nature." Thoreau could not have gotten what he wanted looking at a rose; the rose has been so heavily burdened with human "prejudice, and tradition, and delusion"—with human history—that by now there is no hard bottom to be found there. "The mud and slush of opinion" has been bred right into Dolly Parton; she's more a symptom of culture than a cure for it.

But if Dolly Parton suggests that our intercourse with nature will sometimes produce regrettable offspring, that doesn't necessarily mean we are better off with swamps. It is too late in the day—there are simply too many of us now—to follow Thoreau into the woods, to look to nature to somehow cure or undo culture. As important as it is to have swamps, today it is probably more important to learn how to mingle our art with nature in ways that culminate in a Madame Hardy rather than a Dolly Parton—in forms of human creation that satisfy culture without offending nature. The habit of bluntly opposing nature and culture has only gotten us into trouble, and we won't work ourselves free of this trouble until we have developed a more complicated and supple sense of how we fit into nature. I do not know what that sense might be, but I suspect that the rose, with its long, quirky history of give-and-take with man, can tutor it as well as, if not better than, Thoreau's unsullied swamp.

Even once we have recognized the falseness of the dichotomy between nature and culture, it is hard to break its hold on our minds and our language; look how often I fall back on its terms. Our alienation from nature runs deep. Yet even to speak in terms of a compromise between nature and culture is not quite right either, since it implies a distance between the two—implies that we are not part of nature. So many of our metaphors depend on this rift, on a too-easy sense of what is nature and what is "a color of the spirit." What we need is to confound our metaphors, and the rose can help us do this better than the swamp can.

That perhaps is what matters when we look at a rose blossom and think of sex. In my garden this summer, Maiden's Blush has flowered hugely, some of her blossoms flushed so deeply pink as to deserve the adjective *émue.* So what does it mean to look at these blossoms and think of sex? Am I thinking metaphorically? Well, yes and no. This flower, like all flowers, *is* a sexual organ. The uncultured bumblebee seems to find this bloom just as attractive as I do; he seems just as bowled over by its perfume. Yet I can't believe I gaze at the blossom in quite the same way he does. Its allure, for me, has to do with its resemblance to women—to "the thighs of an aroused nymph," about which I can assume he feels nothing. For this is a resemblance my species has bred, or selected, this rose to have. So is it imaginary? Merely a representation? (But what about the bee?! That's no representation he's pollinating.) Are we, finally, speaking of nature or culture when we speak of a rose (nature) that has been bred (culture) so that its blossoms (nature) make men imagine (culture) the sex of women (nature)?

It may be this sort of confusion that we need more of.

CHAPTER 6

Weeds Are Us

Ralph Waldo Emerson, who as a lifelong gardener really should have known better, once said that a weed is simply a plant whose virtues we haven't yet discovered. "Weed" is not a category of nature but a human construct, a defect of our perception. This kind of attitude, which comes out of an old American strain of romantic thinking about wild nature, can get you into trouble. At least it did me. For I had Emerson's pretty conceit in mind when I planted my first flower bed, and the result was not a pretty thing.

Having read perhaps too much Emerson, and too many of the sort of gardening book that advocates "wild gardens" and nails a pair of knowing quotation marks around the word *weed* (a sure sign of ecological sophistication), I sought to make a flower bed that was as "natural" as possible. Rejecting all geometry (too artificial!), I cut a more or less kidney-shaped bed in the lawn, pulled out the sod, and divided the bare ground into irregular patches that I roughly outlined with a bit of ground limestone. Then I took packets of annual seeds—bachelor's buttons, nasturtiums, nicotianas, cosmos, poppies (California and Shirley both), cleomes, zinnias, and sunflowers—and broadcast a handful of each into the irregular patches, letting the seeds fall wherever nature dictated. No rows: this bed's arrangement would be *natural.* I sprinkled the seeds with loose soil, watered, and waited for them to sprout.

Pigweed sprouted first, though at the time I was so ignorant that I figured this vigorous upstart must be zinnia, or sunflower. I had had no prior acquaintance with pigweed (it grew nowhere else on the property), and did not deduce that it was a weed until I noticed it was coming up in every single one of my irregular patches. Within a week the entire bed was clothed in tough, hairy pigweeds, and it was clear that I would have to start pulling them out if I ever expected to see my intended annuals. The absence of rows or paths made weeding difficult, but I managed to at least thin the lusty pigweeds, and the annuals, grateful for the intervention on their behalf, finally pushed themselves up out of the earth. Finding the coast relatively clear, they started to grow in earnest.

That first summer, my little annual meadow thrived, pretty much conforming to the picture I'd had in mind when I planted it. Sky-blue drifts of bachelor's buttons flowed seamlessly into hot spots thick with hunter-orange and fire-engine poppies, behind which rose great sunflower towers. The nasturtiums poured their sand-dollar leaves into neat, low mounds dabbed with crimson and lemon, and the cleomes worked out their intricate architectures high in the air. Weeding this dense tangle was soon all but impossible, but after the pigweed scare I'd adopted a more or less laissez-faire policy toward the uninvited. The weeds that moved in were ones I was willing to try to live with: jewelweed (a gangly orange-flowered relative of impatiens), foxtail grass, clover, shepherd's purse, inconspicuous Galinsoga, and Queen Anne's lace, the sort of weed Emerson must have had in mind, with its ivory lace flowers (as pretty as anything you might plant) and edible, carrotlike root. That first year a pretty vine also crept in, a refugee from the surrounding lawn. It twined its way up the sunflower stalks and in August unfurled white, trumpet-shaped flowers that resembled morning glory. What right had I to oust this delicate vine? To decide that the flowers I planted were more beautiful than ones the wind had sown? I liked how wild my garden was, how peaceably my cultivars seemed to get along with their wild relatives. And I liked how unneurotic I was being about "weeds." Call me Ecology Boy.

"Weeds," I decided that summer, did indeed have a bad rap. I thought back to my grandfather's garden, to his unenlightened, totalitarian approach toward weeds. Each day he patrolled his pristine rows, beheading the merest smudge of green with his vigilant hoe. Hippies, unions, and weeds: all three made him crazy then, an old man in the late sixties, and all three called forth his reactionary wrath. Perhaps because there was little he could do to stop the march of hippies and organized labor, he attacked weeds all the more zealously. He was one of those gardeners who would pull weeds anywhere—not just in his own or other people's gardens, but in parking lots and storefront window boxes too. His world then was under siege, and weeds to him represented the advance guard of the forces of chaos. Had he lived to see it, my little wild garden—this rowless plant be-in, this horticultural Haight-Ashbury—would probably have broken his heart.

My grandfather wasn't the first person to sense a social or political threat in the growth of weeds. Whenever Shakespeare tells us that "darnel, hemlock, and rank fumatory" or "hateful docks, rough thistles, kecksies, burrs" are growing unchecked, we can assume a monarchy is about to fall. Until the romantics, the hierarchy of plants was generally thought to mirror that of human society. Common people, one writer held in 1700, may be "looked upon as trashy weeds or nettles." J. C. Loudon, an early nineteenth-century gardening expert, invited his readers "to compare plants with men, [to] consider aboriginal species as mere savages, and botanical species . . . as civilized beings."

The garden world even today organizes plants into one great hierarchy. At the top stand the hypercivilized hybrids—think of the rose, "queen of the garden"—and at the bottom are the weeds, the plant world's proletariat, furiously reproducing and threatening to usurp the position of their more refined horticultural betters. Where any given plant falls in this green chain of being has a lot to do with fashion, but there are a few abiding rules. In general, the more intensively a plant has been hybridized—the further it's been distanced from its wildflower origins—the higher its station in plant society. Thus a delphinium can lord it over a larkspur, a heavily doubled bourbon rose over a five-petaled rugosa. A corollary of this rule holds that the more "weedy" a plant is—the easier it is to grow—the lower its place: garden phlox, heir to all the fungi, has greater status than indestructible coreopsis.

Color, too, determines rank, and white comes at the top. This is because pure white occurs only rarely in nature, and perhaps also because a taste for the subtleties of white flowers is something that must be acquired. (Gaudy colors have always been associated with the baser elements: gaillardia, a loud, two-toned cousin of the daisy, used to be called "nigger flower.") Just beneath white is blue, a color that has always enjoyed royal and aristocratic connections, and from there it is a descent downscale through the hot and flashy shades, by the all-too-common yellows, past the reds even bulls will take note of, on down to the very bottom: shunned, rebuffed, eschewed, embarrassing, promiscuous magenta. Magenta, the discount pigment with which nature has brushed a thousand weeds, has always been a mark of bad breeding in the garden world. The offspring of hybrid species that have been allowed to set seed will frequently revert to magenta, as base genes reassert themselves.

The nineteenth-century romantics, who looked more kindly on the common man, also looked kindly on the weed. By the time they wrote, the English countryside had been so thoroughly dominated, every acre cleared of trees and bisected by hedgerows, that the idea of a wild landscape acquired a strong appeal, perhaps for the first time in European history. (Nostalgia for wilderness comes easy once it no longer poses any threat.) Ruskin wrote enthusiastically of the wildflower, which had never been "provoked to glare into any gigantic impudence at a flower show." He judged a flower garden as unnatural, "an ugly thing, even when best managed: it is an assembly of unfortunate beings, pampered and bloated above their natural size; . . . corrupted by evil communication into speckled and inharmonious colours; torn from the soil which they loved, and of which they were the spirit and the glory, to glare away their term of tormented life. . . ."

If garden flowers were slaves to men, then weeds were emblems of freedom and wildness—at least among romantic writers who lived at some distance from nature. "Better to me the meanest weed," wrote Tennyson in the early 1830s. "Weed" soon became a standard synecdoche for *wilderness,* as in this stanza of Gerard Manley Hopkins:

> *What would the world be, once bereft*
> *Of wet and wildness? Let them be left,*
> *O let them be left, wildness and wet;*
> *Long live the weeds and the wilderness yet.*

Predictably, the romance of the weed gained a ready purchase on the American mind, which has always been disposed to regard the works of nature as superior to those of men, and to resist hierarchies wherever they might be found. The weed supplies Emerson, Whitman, Thoreau, and generations of American naturalists with a favorite trope—for unfettered wildness, for the beauty of the unimproved landscape, and of course, when in quotes, for the benightedness of those fellow countrymen who fail to perceive nature as acutely and sympathetically as they do. (We'll leave aside for now the question of how acutely these writers themselves perceived weeds.) Weed worship continues to flower periodically in America, most recently in the sixties. "Weed" became a fond nickname for marijuana, and millions of us consulted our tattered copies of Euell Gibbons's *Stalking the Wild Asparagus,* an improbable best-seller that, essentially, proposed weeds as the basis of a new American cuisine. Whenever history and culture seem stifling, weeds begin to look good.

My own romance of the weed did not survive a second summer. The annuals, which I had allowed to set seed the previous year, did come back, but they proved a poor match for the weeds, who returned heavily reinforced. It was as though news of this sweet deal (this chump gardener!) had spread through the neighborhood over the winter, for the weed population burgeoned, both in number and in kind. Recognizing that what I now tended was a weed garden, and having been taught that a gardener should know the name of every plant in his care, I consulted a few field guides and drew up an inventory of my collection. In addition to the species I've already mentioned, I had milkweed, pokeweed, smartweed, St.-John's-wort, quack grass, crabgrass, plantain, dandelion, bladder campion, fleabane, butter-and-eggs, timothy, mallow, bird's-foot trefoil, lamb's quarters, chickweed, purslane, curly dock, goldenrod, sheep sorrel, burdock, Canada thistle, and stinging nettle. I'm sure I've missed another dozen, and misidentified a few, but this will give you an idea of the various fruits of my romanticism. What had begun as a kind of idealized wildflower meadow now looked like a roadside tangle and, if I let it go another year, would probably pass for a vacant lot.

Since this had not been my aesthetic aim, I set about reclaiming my garden—to at least arrest the process at "country roadside" before it degenerated to "abandoned railroad siding." I would be enlightened about it, though, pardoning the weeds I liked and expelling all the rest. I was prepared to tolerate the fleabane, holding aloft their sunny clouds of tiny asterlike flowers, or milkweed, with its interesting seedpods, but bully weeds like burdock, Canada thistle, and stinging nettle had to go. Unfortunately, the weeds I liked least proved to be the best armed and most recalcitrant. Burdock, whose giant clubfoot leaves shade out every other plants for yards around, holds the earth in a death grip. Straining to pull out its mile-long taproot, you feel like a boy trying to arm-wrestle a man. Inevitably the root breaks before it yields, with the result that, in a few days' time, you have two tough burdocks where before there had been one. All I seemed able to do was help my burdock reproduce. I felt less like an exterminator of these weeds than their midwife.

That pretty vine with the morning glory blossoms turned out to be another hydra-headed monster. Bindweed, as it's called, grows like kudzu and soon threatened to blanket the entire garden. It can grow only a foot or so high without support, so it casts about like a blind man, lurching this way then that, until it finds a suitable plant to lean on and eventually smother. Here too my efforts at eradication proved counterproductive.

Bindweed, whose roots may reach ten feet down, can reproduce either by seed or human-aided cloning. For its root is as brittle as a fresh snapbean; put a hoe to it and it breaks into a dozen pieces, *each of which will sprout an entire new plant.* It is as though the bindweed's evolution took the hoe into account. By attacking it at its root—the approved strategy for eradicating most weeds—I played right into the insidious bindweed's strategy for world domination.

Have I mentioned my annuals? A few managed to hang on gamely. California poppies and Johnny-jump-ups proved adept at finding niches among the thistles, and a handful of second-generation nicotianas appeared, though these had reverted to the hue of some weedy ancestor—instead of bright pink, they came back a muddy shade of pale green. For the most part, my annuals counted themselves lucky to serve as underplanting for the triumphant weeds. But whatever niches remained for them the grasses seemed bent on erasing. Stealthy quack grass moved in, spreading its intrepid rhizomes to every corner of the bed. Quack grass roots can travel laterally as much as fifty feet, moving an inch or two beneath the surface and pushing up a blade (or ten) wherever the opportunity arises. You pull a handful of this grass thinking you've doomed an isolated tuft, only to find you've grabbed hold of a rope that reaches clear into the next county—where it is no doubt tied by a very good knot to an oak.

Now what would Emerson have to say? I had given all my weeds the benefit of the doubt, acknowledged their virtues and allotted them a place. I had treated them, in other words, as garden plants. But they did not behave as garden plants. They differed from my cultivated varieties not merely by a factor of human esteem. No, they seemed truly a different order of being, more versatile, better equipped, swifter, craftier—simply more adroit at the work of being a plant. What garden plant can germinate in thirty-six minutes, as a tumbleweed can? What cultivar can produce four hundred thousand seeds on a single flower stalk, as the mullein does? Or hitch its seeds to any passing animal, like the burdock? Or travel a foot each day, as kudzu can? ("You keep still enough, watch close enough," southerners will tell you, "and damn if you can't see it move.") Or, like the bindweed, clone new editions of itself in direct proportion to the effort we expend trying to eradicate it? Japanese knotweed can penetrate four inches of asphalt, no problem. Each summer the roots of a Canada thistle venture another ten feet in every direction. Lamb's-quarter seeds recovered from an archaeological site germinated after spending seventeen hundred years in storage, patiently awaiting their shot. The roots of the witchweed emit a poison that kills every other plant in its vicinity.

No, it can't just be my lack of imagination that gives the nettle its sting.

So what is a weed? I consulted several field guides and botany books hoping to find a workable definition. Instead of one, however, I found dozens, though almost all of them could be divided into two main camps. "A weed is any plant in the wrong place" fairly summarizes the first camp, and the second maintains, essentially, that "a weed is an especially aggressive plant that competes successfully against cultivated plants." In the first, Emersonian definition, the weed is a human construct; in the second, weeds possess certain inherent traits we did not impose. The metaphysical problem of weeds, I was beginning to think, is not unlike the metaphysical problem of evil: Is it an abiding property of the universe, or an invention of humanity?

Weeds, I'm convinced, are really out there. But I am prepared to concede the existence of a gray area inhabited by Emerson's weeds, plants upon which we have imposed weediness simply because we can find no utility or beauty in them. One man's flowers may indeed be another's weeds. Purple loosestrife, which I planted in my perennial border, has been outlawed as a "noxious weed" in several midwestern states, where it has escaped gardens and now threatens wetland flora. Likewise, certain of my weeds may have value in the eyes of another. Every day I pull easily enough dandelions and purslanes from my vegetable garden to make a tasty salad for Euell Gibbons. What I call weeds he would call lunch.

Not long ago, I had a local excavator over to estimate a job for me. He was one of those venerable old-timers who possess an intimate knowledge of local geography. This fellow knew more about my land than I did: how many gallons a minute my well could pump, and the source of its water; the alkalinity of my soil, due to the limestone ledge it sits on; the fact that the ancient apple trees on my property used to make the best hard cider in town. We walked around the property, looking for a suitable spot for a pond, and he nodded approvingly at the changes I'd made in the landscape: restoring a meadow that had gone to brush, pruning the old apple trees, regrading the hillside to divert spring runoff from the house. But there was one thing I'd done that seemed to bother him, and after a while he spoke up. I'd planted a pair of weeping willows at the edge of a small wetland. The size of phone booths when I put them in, two years later they were already as big as houses. The property has few shade trees—it *looks* hot in the summer—and I planted the willows to create a few cool spots in the landscape; willows seem almost to imply water, and they will amplify the quietest breeze. My visitor tipped his head in the direction of the trees and growled:

"What'n hell did you plant those weeds for?"

"Weeds?! What are you talking about?" The willows were poetry as far as I was concerned.

"Those damn trees. They're good for nothing, they're dirty, and if you don't watch it, their roots are going to crack your foundation one day. You'll see."

I've since discovered that a lot of country people consider willows weeds. Given a good supply of water, their growth is rampant, and their roots have been known to bust through concrete. But my trees were a good fifty yards from the house. The real objection, it seems, is that willows drop a lot of branches in the course of a growing season; in the eyes of a dedicated lawn man, they *are* dirty. They're also faulted for having soft wood, the result of their rapid growth. The world of trees has its own hierarchy, of course, and hardwood stands at the top. Soft, wet woods like willow have no commercial value, either as lumber or firewood.

So a perspective exists from which the weeping willow is a weed. It grows too fast, sullies lawns, damages homes, and burns about as well as celery. It also grows wild around here, another supposedly weedy trait. My excavator's dealings with willows were mainly a matter of economics; he knew you could not sell a cord of willow to save your life, and he'd heard enough stories about cracked concrete. My own dealings were largely aesthetic, and they were informed by another kind of story, one that put the willow in a completely different light.

The weeping willow, I had read, is not a native tree, but an eighteenth-century garden import. It is thought to have first been planted in America not too far from here, in the Stamford, Connecticut, garden of Samuel Johnson, a clergyman and philosopher who was the first president of Kings College (which was renamed Columbia University after independence). Johnson saw his first weeping willow at Twickenham, Alexander Pope's famous garden on the Thames. He was so taken with Pope's ancient tree that he returned to his home on the Housatonic with a cutting. Evidently the banks of that river proved as hospitable to the tree as those of the Thames, because the weeping willow soon escaped his garden and spread north. Today vast, pendulous willows—towering green fountains—line the Housatonic from Stamford to the Berkshires; every one of them, presumably, can trace its lineage back to Pope's great garden. Knowing this, I am more apt to think of Twickenham and the Thames when I see a weeping willow than I am of a cement-busting "weed."

These stories about weeping willows would seem to bolster Emerson's contention that weediness is in the eye of the beholder, that it is a matter of perception. Now ordinarily I am perfectly comfortable with this sort of relativistic thinking, but experience suggests that here it is shallow.

And not only my experience: Emerson's own disciple, Henry David Thoreau, runs into some difficulty with his teacher's theory of weeds when he plants his bean field at Walden.

As an observer and naturalist, Thoreau consistently refuses to make "invidious distinctions" between different orders of nature; sworn enemy of hierarchy, the man boasts of the fact that he loves swamps more than gardens. But as soon as he determines to make "the earth say beans instead of grass" he finds he has made enemies in nature: worms, the morning dew, woodchucks, and weeds. The bean field "attached me to the earth," Thoreau felt, giving him positions he must defend if he hopes to prove his experiment in self-reliance a success. And so Thoreau is obliged to wage a long and decidedly uncharacteristic "war, not with cranes, but with weeds, those Trojans who had sun and rain and dews on their side. Daily the beans saw me come to their rescue armed with a hoe, and thin the ranks of their enemies, filling up the trenches with weedy dead." He finds himself "making such invidious distinctions with his hoe, levelling whole ranks of one species, and sedulously cultivating another."

Thoreau is gardening here, of course, and this forces him at least for a time to throw out his romanticism about nature—to drop what naturalists today hail as his precocious "biocentrism" (as opposed to anthropocentrism). But by the end of the chapter, his bean field having achieved its purpose, Thoreau trudges back—lamely, it seems to me—to the Emersonian fold: "The sun looks on our cultivated fields and on the prairies and forests without distinction. . . . Do [these beans] not grow for woodchucks too? . . . How, then, can our harvest fail? Shall I not rejoice also at the abundance of the weeds whose seeds are the granary of the birds?"

Sure, Henry, rejoice. And starve.

My own experience in the garden has convinced me "absolute weediness" does exist—that weeds represent a different order of being, and the fact that Thoreau's beans were no match for his weeds does not mean the weeds have a higher claim to the earth, as Thoreau seems to think. I found support for this hunch in the field guides and botany books I consulted when I was trying to identify my weeds. As I searched these volumes for the *noms de bloom* of my marauders, I jotted down each species' preferred habitat. Here are a few of the most typical: "waste places and roadside"; "open sites"; "old fields, waste places"; "cultivated and waste ground"; "old fields, roadsides, lawns, gardens"; "lawns, gardens, disturbed sites."

What this list suggests is that weeds are not superplants: they don't grow everywhere, which explains why, for all their vigor, they haven't covered the globe entirely. Weeds, as the field guides indicate, are plants particularly well adapted to man-made places. They don't grow in forests or prairies—in "the wild." Weeds thrive in gardens, meadows, lawns, vacant lots, railroad sidings, hard by dumpsters and in the cracks of sidewalks. They grow where we live, in other words, and hardly anywhere else.

Weeds, contrary to what the romantics assumed, are not wild. They are as much a product of cultivation as the hybrid tea rose, or Thoreau's bean plants. They do better than garden plants for the simple reason that they are better adapted to life in a garden. For where garden plants have been bred for a variety of traits (tastiness, nutritiousness, size, aesthetic appeal), weeds have evolved with just one end in view: the ability to thrive in ground that man has disturbed. At this they are very accomplished indeed.

Weeds stand at the forefront of evolution; no doubt they are evolving in my garden at this very moment, their billions of offspring self-selecting for new tactics to outwit my efforts and capitalize on any opening in my garden. Weeds are nature's ambulance chasers, carpetbaggers, and confidence men. Virtually every crop in general cultivation has its weed impostor, a kind of botanical doppelganger that has evolved to mimic the appearance as well as the growth rate of the cultivated crop and so ensure its survival. Some of these impostors, such as wild oats, are so versatile that they can alter their appearance depending on the crop they are imitating, like an insidious agricultural Fifth Column. According to Sara B. Stein's botany, *My Weeds*, wild oats growing in a field of alternating rows of spring and winter barley will mimic the habits of either crop, *depending on the row.* Stein, whose book is a trove of information about weeds, also tells of a rice mimic that became so troublesome that researchers planted a purple variety of rice to expose the weeds once and for all. Within a few years, the weed-rice had turned purple too.

And yet as resourceful and aggressive as weeds may be, they cannot survive without us any more than a garden plant can. Without man to create crop land and lawns and vacant lots, most weeds would soon vanish. Bindweed, which seems so formidable in the field and garden, can grow nowhere else. It lives by the plow as much as we do.

To learn all this was somehow liberating. My weeds were no more natural than my garden plants, had no greater claim to the space they were vying for. Those smug quotes in which naturalists like to coddle weeds were merely a conceit. My battles with weeds did not bespeak alienation from nature, or some irresponsible drive to dominate it. Had Thoreau known this, perhaps he would not have troubled himself so about "what right had I to oust St. Johnswort, and the rest, and to break up their ancient herb garden?"

Thoreau considered his wormwood, pigweed, sorrel, and St.-John's-wort part of nature, his beans part of civilization. He looked to the American landscape, as many of us do, for a path that would lead him out of history and into nature, and this led him to value what grew "naturally" over that which man planted. But as it turns out history is inescapable, even at Walden. Much of the flora in the Walden landscape is as historical as his beans, his books, even the Mexican battlefield he makes his bean field a foil for. Had Thoreau brought a field guide with him to Walden, he might have noted that most of the weeds that came up in his garden were alien species, brought to America by the colonists. St.-John's-wort, far from being an ancient Walden resident, was brought to America in 1696 by a band of fanatical Rosicrucians who claimed the herb had the power to exorcise evil spirits. You want to privilege *this* over beans?

It's hard to imagine the American landscape without St.-John's-wort, daisies, dandelions, crabgrass, timothy, clover, pigweed, lamb's-quarters, buttercup, mullein, Queen Anne's lace, plantain, or yarrow, but not one of these species grew here before the Puritans landed. America in fact had few indigenous weeds, for the simple reason that it had little disturbed ground. The Indians lived so lightly on the land that they created few habitats for weeds to take hold in. No plow, no bindweed. But by as early as 1663, when John Josselyn compiled a list "of such plantes as have sprung up since the English planted and kept cattle in New England," he found, among others, couch grass, dandelion, sow thistle, shepherd's purse, groundsel, dock, mullein, plantain, and chickweed.

Some of these weeds were brought over deliberately: the colonists prized dandelion as a salad green, and used plantain (which is millet) to make bread. Other weed seeds, though, came by accident—in forage, in the earth used for shipboard ballast, even in pants cuffs and cracked boot soles. Once here, the weeds spread like wildfire. According to Alfred W. Crosby, the ecological historian, the Indians considered the Englishman a botanical Midas, able to change the flora with his touch; they called plantain "Englishman's foot" because it seemed to spring up wherever the white man stepped. (Hiawatha claimed that the spread of the plant presaged the doom of the wilderness.) Though most weeds traveled with white men, some, like the dandelion, raced west of their own accord (or possibly with the help of the Indians, who quickly discovered the plant's virtues), arriving well ahead of the pioneers. Thus the supposedly virgin landscape upon which the westward settlers gazed had already been marked by their civilization. However, those same pioneers did *not* gaze out on tumbleweed, that familiar emblem of the untamed western landscape. Tumbleweed did not arrive in America until the 1870s, when a group of Russian immigrants settled in Bonhomme County, South Dakota, intending to grow flax. Mixed in with their flax seeds were a few seeds of a weed well known on the steppes of the Ukraine: tumbleweed.

European weeds thrived here, in a matter of years changing the face of the American landscape, helping to create what we now take to be our country's abiding "nature." Why should these species have prospered so? Probably because the Europeans who brought them got busy making the land safe for weeds by razing the forests, plowing fields, burning prairies, and keeping grazing animals. And just as the Europeans helped smooth the way for their weeds, weeds helped smooth the way for Europeans. This is particularly true in the case of the grasses. Native grasses proved poor forage for European livestock, which at first did not fare well in America. Yet colonists noted that after a few years the grasses—and in turn the health of the livestock—seemed to improve. What had happened, according to Crosby, is that Old World livestock had overgrazed the native grasses. Because these species were unaccustomed to such heavy grazing, they had trouble regenerating themselves. This left them vulnerable to the onslaught of European weed grasses which, having co-evolved with the goat and sheep and cow, are better equipped to withstand the grazing pressure of these animals. The European grasses soon conquered American meadows, thereby providing European livestock with their preferred forage once again. Today most of the native grasses have vanished.

Working in concert, European weeds and European humans proved formidable ecological imperialists, rapidly driving out native species and altering the land to suit themselves. The new plant species thrived because they were consummate cosmopolitans, opportunists superbly adapted to travel and change. In a sense, the invading species had less in common with the retiring, provincial plants they ousted than with the Europeans themselves. Or perhaps that should be put the other way around. "If we confine the concept of weeds to species adapted to human disturbance," writes Jack R. Harland in *Crops and Man,* "then man is by definition the first and primary weed under whose influence all other weeds have evolved."

Weeds are not the Other. Weeds are us.

A pedestrian standing at the corner of Houston Street and LaGuardia Place in Manhattan might think that the wilderness had reclaimed a tiny corner of the city's grid here. Ten years ago, an environmental artist persuaded the city to allow him to create on this site a "time landscape" showing New Yorkers what Manhattan looked like before the white man arrived. On a small hummock he planted oak, hickory, maples, junipers, and sassafras, and they've grown up to form a nearly impenetrable tangle, which is protected from New Yorkers by the steel bars of a fence now thickly embroidered with vines. It's exactly the sort of "garden" of which Emerson and Thoreau would have approved—for the very reason that it's *not* a garden. Or at least that's the conceit.

I walk by this anti-garden most mornings on my way to work, and for some reason it has always irritated me. It adjoins a lively community garden, where any summer evening will find a handful of neighborhood people busy cultivating their little patches of flowers and vegetables. Next to this display of enterprise, the untended "time landscape" makes an interesting foil. But the juxtaposition has always struck me as pat, just a shade too righteous, and walking by one day last summer I figured out why.

My mind fixed on the weeds just then hoisting their flags of victory over my garden, I recognized one of the vines twining along the fence from the field guides I'd been consulting. It was nightshade, a species, I recalled—and not without my own sweet pang of righteousness—that is not indigenous: it came to America with the white man. Aha! This smug little wilderness was really a garden after all. Unless somebody weeds it, sedulously and knowledgeably, it will quickly be overrun with alien species. This "time landscape" is in perpetual danger of degenerating into an everyday vacant lot; only a gardener, armed with a hoe and a set of "invidious distinctions," can save it.

Once, of course, this would not have been the case. But that was a long time ago; by now, we have made so many changes in the land that some form of gardening has become unavoidable, even in those places we wish to preserve as monuments to our absence. This, it seems to me, is one of the lessons of the massive fires in Yellowstone in 1988. At a certain point in history, doing nothing is not necessarily benign. Since 1972, park management in Yellowstone has followed a policy called "natural burn," under which naturally occurring fires are allowed to burn freely—before 1972, every fire was put out immediately. All those years of fire fighting left an abundance of volatile dead wood on the forest floor and that may be why, when the fires finally came in the drought year of 1988, they proved so catastrophic. Yellowstone's ecosystem having already been altered by the earlier policy of fire suppression, the new policy could not in any real sense be "natural"; nor were the fires it fostered.

There's no going back. Even Yellowstone, our country's greatest "wilderness," stands in need of careful management—it's too late to simply "leave it alone." I have no idea what the best fire policy for Yellowstone might be, but I do know that men and women, armed with scientific knowledge and acting through human institutions, will have to choose and then implement one. In doing so, they will have to grapple with the fact that, long before Yellowstone was declared a "wilderness area," Indians were setting fires in it; were these "natural"? If the goal is to restore Yellowstone to its pre-Columbian condition, their policy may well have to include the setting of fires. They will also have to decide how many tourists Yellowstone can support, whether wolves should be reintroduced to keep the elk population from exploding, and a host of other complicated questions. Today, even Yellowstone must be "gardened."

A century after Thoreau wrote that "in wildness is the preservation of the world," Wendell Berry, the Kentucky poet and farmer, added a corollary that would have made no sense at all to Thoreau, and yet that is necessary. Berry wrote that "in human culture is the preservation of wildness." I take him to mean that it's too late now to do nothing. Only human wisdom and forbearance can save places like Yellowstone.

Thoreau, and his many heirs among contemporary naturalists and radical environmentalists, assume that human culture is the problem, not the solution. So they urge us to shed our anthropocentrism and learn to live among other species as equals. This sounds like a fine, ecological idea, until you realize that the earth would be even worse off if we started behaving any more like animals than we already do. The survival strategy of most species is to extend their dominion as far and as brutally as they can, until they run up against some equally brutal natural limit that checks their progress. Isn't this exactly what we've been doing?

What sets us apart from other species is culture, and what is culture but forbearance? Conscience, ethical choice, memory, discrimination: it is these very human and decidedly unecological faculties that offer the planet its last best hope. It is true that, historically, we've concentrated on exercising these faculties in the human rather than the natural estate, but that doesn't mean they *cannot* be exercised there. Indeed, this is the work that now needs to be done: to bring more culture to our conduct in nature, not less.

If I seem to have wandered far afield of weeds, consider what weeding is: the process by which we make informed choices in nature, discriminate between good and bad, apply our intelligence and sweat to the earth. To weed is to bring culture to nature—which is why we say, when we are weeding, that we are *cultivating* the soil. Weeding, in this sense, is not a nuisance that follows from gardening, but its very essence. And, like gardening, weeding at a certain point becomes an obligation. As I learned in my flower bed, mere neglect won't bring back "nature."

In this, my yard is not so different from the rest of the world. We cannot live in it without changing nature irrevocably; having done so, we're obliged to tend to the consequences of the changes we've wrought, which is to say, to weed. "Weeding" is what will save places like Yellowstone, but only if we recognize that weeding is not just something we do to the land—only if we recognize the need to cultivate our *own* nature, too. For though we may be the earth's gardeners, we are also its weeds. And we won't get anywhere until we come to terms with this crucial ambiguity about our role—that we are at once the problem and the only possible solution to the problem.

Eventually I came to see that my weed-choked natural garden was irresponsible. My garden plants had thrown in their lot with me, and I had failed to protect them from the weeds. So I ripped out my flower garden and began anew. This time, I cut a perfect rectangle in the grass, and planted my flower seeds in scrupulous rows, eighteen inches apart and as straight as a plumb line could make them. As the seedlings came up, I cultivated assiduously between the rows, using the Dutch hoe that my grandfather had given me. I didn't worry much about epistemology: whatever came up between the rows I judged a weed and cut it down. The rows began as a convenience—they make cultivation an easy matter—but I've actually come to like the way they look; I guess by now I am more turned off by romantic conceits about nature than by a bit of artifice in the garden. Geometry is man's language, Le Corbusier once said, and I am happy to have a garden that speaks in that tongue. I know better now than to think a less tended garden is any more natural; weeds are our words, too.

As I see it, the day I decided to disturb the soil, I undertook an obligation to weed. For this soil is not virgin and hasn't been for centuries. It teems with hundreds of thousands of weed seeds for whom the thrust of my spade represents the knock of opportunity. Not "nature," strictly speaking, these seeds are really the descendants of earlier gardeners. To let them grow, to do nothing, is tantamount to letting those gardeners plant my garden: to letting all those superstitious Rosicrucians and Puritans and Russian immigrants have their way here. To do nothing, in other words, would be no favor to me, or my plants, or to nature. So, I weed.

CHAPTER 7

Green Thumb

I am alleged to have one, at least by those of my friends who garden with less success than I do. That's usually how it goes with a green thumb—nobody quite believes it of themselves, but when you find someone whose beefsteaks are fat and red by July, and whose delphiniums soar like periwinkle skyscrapers over the prosperous city of their perennial border, the term fairly leaps to the tongue. It figures: Your own failures will seem more bearable if the other gardener has a gift from the gods.

Though I am reasonably sure there is such a thing, I'm not about to number myself among the graced and elect. It would be, I don't know, a bit presumptuous. Maybe even dangerous—like I was really asking for it. (How about a blast of August frost, Mister Green Thumb? Or maybe a plague of aphids?) I guess I'm a little like the Calvinist who doesn't dare assume anything about his status, gracewise. And even though he knows it's probably already settled, he's either got it or he doesn't, he's going to keep working at it anyway, just to be on the safe side, cover all the bases. Besides, by now my alleged green thumb (Let *them* say it!) is a reputation I feel obliged to at least try to uphold. So I fret over my transplants, monitor my soil closely, thumb through the reference books—as if good works in the garden might be taken for grace. All of which makes my many failures that much harder to bear.

Consider my carrots. Every spring I planted them, and every summer I pulled from my soil this very sorry collection of gnarled, arthritic digits—all knuckle, and not one more than two inches long. I might have learned to accept this gap in my gardening repertoire except for the fact that carrots are generally regarded as one of the easier vegetables to grow. They come up readily from seed, few pests have a taste for them, and they're untroubled by frost. Buy one of those "child's first garden" kits and it's bound to contain a packet of carrot seeds. That's not only because carrots figure prominently in the childhood imagination (think of Bugs Bunny, Captain Kangaroo), but because they're considered more or less "foolproof."

What sort of green thumb could I possibly have if I couldn't grow a carrot? This failure was an embarrassment, frankly, and a crisis of my gardening faith.

So I determined to get it right, to know carrots. I thought long and hard about them. I even tried to think what my carrots might be thinking—to imagine what it was about their situation they didn't like. Their tops were lush and green, so their complaint was not about the food or water. Could it be the company? One year they occupied a plot next to the onions, a dubious neighborhood for any plant. (Onions are as controversial in the society of plants as in our own; many species recoil from them.) So the next season I relocated them to a spot by a row of comparatively genial lettuces—and observed no improvement.

What does a carrot care about? This is not as dumb a question as it sounds. It is more than an anthropomorphic conceit to attribute likes and dislikes to plants, to wonder, if not about how they're "feeling," then at least about what matters to them, what they require in order to fulfill the terms of their destiny. Most of the good gardeners I've met seem to possess a faculty, akin to empathy, that allows them to sense what their plants might need at any time. "If you wish to make anything grow," Russell Page wrote in *The Education of a Gardener,* "you must understand it, and understand it in a very real sense. 'Green fingers' are a fact, and a mystery only to the unpracticed. But green fingers are the extensions of a verdant heart." I don't think Page is merely being sentimental here. He's speaking of the need in gardening for an imaginative leap—in my case, into the innermost nature of carrothood. And this is what I attempted. I considered, What would matter most to a carrot as it struggled to get past the pinkie stage? And it came to me: shoulder room.

I pictured a cross section of the first few inches of my soil and it was a Number 6 train at rush hour, jammed with cramped orange commuters. My carrots stood too close together; I had been insufficiently ruthless when it came to thinning the seedlings. (This seems to be a common failing among inexperienced gardeners; killing off that which you've just planted seems wasteful, even cruel, but triage is essential in the case of root crops.) And I imagined something else, too: that a carrot, aspiring to drive its taproot straight down into the earth, would want an airy soil, no hard clumps or stones to impede its thrust. Had I given my carrots such a root run? It was easy to find out. I stuck my index finger into the soil and barely reached the second knuckle before jamming up against thick, wet clay. My soil was too heavy for carrots.

Success, they say, is a matter of being in the right place at the right time. Perhaps more than is the case in life, in the garden you can often alter the place (and indeed sometimes even the time). I set about giving my carrots a more propitious place by lightening the soil in which they grew. As early in the spring as the ground could be worked, I dug in a bag of builder's sand, a bale of peat moss, and as much compost as I could spare. Ordinarily, carrots wouldn't warrant such a heavy investment of compost, but much was at stake, and nothing works better to lighten a clay soil. I mixed everything together by hand, taking care to remove stones and crush clay clumps as I went along. It took only a few moments of kneading before the consistency of the soil, formerly as dense as fudge, lightened to that of cake. I stuck my finger in again, and sunk without effort to the depth of a cigar. Here was carrot utopia.

After raking the area level and smooth, I sowed two rows of Mokum, a reputedly extra-sweet snub-nosed carrot from France. In a week there emerged a feathery strip of seedlings which I scrupulously thinned to an interval of one inch; a month after, I thinned it again to make absolutely sure my carrots would never have to jostle one another. Everybody now had a seat, and by August I was pulling out of the ground long, orange panatelas, some of the handsomest carrots, I don't mind saying, that I had ever seen. Harvesting root crops has to be one of gardening's finer pleasures. There's the element of surprise (until now, you could only infer by foliage what might be going on down there), and, even better, the small miracle of finding form and color and value amid the earth's black and undifferentiated mass. It's gold prospecting writ small, and these carrots represented quite a strike. I wiped one clean on my shirt, buffed it bright, and then tasted its cool, subterranean sweetness, its unexpectedly intense . . . carrotness.

Maybe, I thought to myself, maybe I did have a green thumb.

Of course a real green thumb would have done everything I did without having to think about it so much. But you have to start somewhere. The seamless, unforced play of a concert pianist starts with a kid picking out tunes on the piano, right? His fingers didn't always know what to do. It's the same with riding a bicycle, or fathoming carrotness. Only much later does it become second nature. Now, I get it—indeed, can no longer imagine *not* getting it—and from here on I'll probably grow fine carrots without a moment's reflection, no bigger a deal than riding a bike. So maybe that is what a green thumb is, a particular form of memory: a compendium of little stories that have been distilled to the point where the gardener can draw on their lessons without even thinking about it—the morals of these stories (most of which are about his own experiences, but some of which may be secondhand) are always at his fingertips.

Most of these narratives would be, like the chronicle of my carrots, tales of failure overcome. All the accomplished gardeners I know are surprisingly comfortable with failure. They may not be happy about it, but instead of reacting with anger or frustration, they seem freshly intrigued by the peony that, after years of being taken for granted, suddenly fails to bloom. They understand that, in the garden at least, failure speaks louder than success. By that I don't mean the gardener encounters *more* failure than success (though in some years he will), only that his failures have more to say to him—about his soil, the weather, the predilections of local pests, the character of his land. The gardener learns nothing when his carrots thrive, unless that success is won against a background of prior disappointment. Outright success is dumb, disaster frequently eloquent. At least to the gardener who learns how to listen.

The book on my soil—which is perhaps the most important volume of information in any gardener's possession—has been written in this language of failure. From my stumpy carrots I learned that it was heavy, consisting of more clay than loam, and acting on that tip I've worked hard to lighten it. When my first crop of tomatoes came up all leaf and scant fruit I realized I had overfertilized—that there had been more nitrogen in the soil to begin with than I'd expected (possibly because my vegetable garden occupies the site of a former cow pasture). The prompt death of a blueberry bush I planted soon after moving here was my first inkling that the soil was sweeter than average for New England. I read that blueberries demand an acid soil, so I wrote off the loss of this bush as a kind of pricey pH test and began each spring to add as much peat moss (which is acidic) as I could afford. Failure sends the gardener to the library, and he returns with a subtler sense of his landscape.

Maybe what I'm describing sounds less like the way of a green thumb than that of any good student or keen observer. Certainly there are many gardeners who hold that the whole green thumb concept is a mystification perpetrated by third-rate gardeners and novices. Russell Page, admittedly a bit of a mystic, is pretty much out there by himself with his verdant heart and green digits. More representative of the gardening world's line on the issue is Eleanor Perényi, who will have none of this green thumb nonsense: "People who blame their failures on 'not having a green thumb' (and they are legion) usually haven't done their homework. There is of course no such thing as a green thumb. Gardening is a vocation like any other—a calling, if you like, but not a gift from heaven. One acquires the necessary skills and knowledge to do it successfully, or one doesn't."

I don't know. Perényi reminds me of a biology teacher I had in the eighth grade, another dutiful demystifier, inveterate empiricist, and wearer of sensible shoes. First class of the year, Mrs. Voigt announced, in a smug tone of voice striving for the matter-of-fact, that a human being was nothing more than a collection of chemicals that could be had from a biological supply company for approximately four dollars. Why so cheap? Because we were 95 percent water, with the rest consisting of relatively common forms of carbon. I knew that day that, even if Mrs. Voigt was right, she was not going to teach me anything I needed to know.

Everything that lives is 95 percent water. Genius is 95 percent perspiration, 5 percent inspiration. Success is 95 percent hard work. Okay, I get it, *but what about that 5 percent?* Tell me watermelon is *99* percent water and you still haven't told me anything interesting—like, what about the 1 percent? Because chances are that's where you're going to find the watermelon.

Perényi's right, as far as she goes. A good gardener observes, remembers, consults the big tomes. But I know I've had occasion to observe something else in good gardeners, a certain touch, an empathy for their plants, a sense of their soil more subtle and complete than any lab report's. There are things they know I can't find in books. It's the difference between the well-trained musician and the maestro, the water and the watermelon. It's that unaccountable 5 percent.

Before we can approach the mysteries of success in the garden, it would help to know something more of failure, a subject on which I can speak with a good deal more authority. I count three main types of failure in the garden. The first is the most straightforward: what both ministers and insurance agents refer to as "acts of God." I'm talking about frosts in August, seventeen-year locusts, droughts, floods, all those Old Testament–type events the gardener can only gape at. Dramatic as these reversals can be, they actually have little to teach the gardener about his garden. (They have rather more to say about the big questions of existence, but these are beyond the scope of this chapter.) The greenest thumb would have helped Pharaoh's gardener not a whit.

The more common varieties of garden failure also originate in nature, but they are more susceptible to our efforts. These I divide into failures of under- and overcultivation.

Some failures of undercultivation would be the predawn raid on your seedlings by a gang of marauding woodchucks; stumpy carrots; a flower bed choked by weeds; a clematis that refuses to bloom; tomatoes that fail to ripen before frost. Failures of undercultivation usually indicate that the gardener has been reluctant to alter the landscape to the extent his plants require; he has not sufficiently tamed nature. Perhaps because of his romantic notions about animals or weeds, he didn't do enough to protect his plants from their incursions. Or he assumed the soil in its unimproved state was adequate to the needs of his trees or tomatoes. Much of this book, in fact, has been an account of my own numerous failures of undercultivation. This is probably because I approached the garden from the direction of the city, and started out with many naïve, urban notions about nature. I assumed I could make a garden and at the same time remain on warm terms with the local flora and fauna. The process of overcoming my failures taught me how much harder it is to get along with nature as an active participant than as a distant admirer. From my failures I learned to be less afraid to exercise human power in nature, to do what is necessary to make the land conform to our designs and supply our needs.

Of course the gardener can push nature too far, and when that happens, he is prone to the third type of failure: failures of overcultivation. The gardener who uses large quantities of fertilizer to coax quick growth from his plants will find them more susceptible to insects and disease. If he adopts an inflexible line on insects, he's apt to spray so much pesticide that he deadens his soil; the bugs are gone, but suddenly nothing seems to grow very well. Plants remain healthy only to the extent they are wild—"able to collaborate with earth, air, light, and water in the way common to plants before humans walked the earth," in Wendell Berry's sensible definition. When cultivation is too intensive it compromises wildness and thereby courts failure.

The blame for some failures of overcultivation should be shared. If a rose succumbs to black spot despite your best efforts, the fault may well lie not with you but the hybridizers. In their quest for improbably large and exotically colored roses, they've created varieties so overbred—so far removed from their wild ancestors—that they are practically guaranteed to fail. (Of course it's up to the gardener to realize that a blue rose may be pushing nature too far; by planting one, he's probably asking for it.) Similarly, the gardener who persists in planting witch hazel up in zone four is pushing it; nature will not abide this sort of hubris, at least not indefinitely.

If I am right to see Scylla and Charybdis on either side of the garden, and if failure is the result of leaning too far in the direction of one or the other, then maybe the green thumb's gift is his sense of balance. The green thumb is the gardener who can nimbly walk the line between the dangers of over- and undercultivation, between pushing nature too far and giving her too much ground. His garden is a place where her ways and his designs are brought gracefully into alignment. To occupy such a middle ground is not easy—the temptation is always to either take complete control or relinquish it altogether, to invoke your own considerable (but in the end overrated) power or to bend to nature's. The first way is that of the developer, the second that of the "nature lover." The green thumb, who will be neither heroic nor romantic, avoids both extremes. He does not try to make water run uphill, but neither does he let it flow wherever it will.

As a metaphor for someone adept at this middle way "green thumb" strikes me as exactly right, marrying as it does natural and human power. Greenness is that force of nature the gardener strives to channel or harness (if not master), and he accomplishes this only to the extent that he understands it deeply, imaginatively. The green thumb fathoms carrotness, can think like water. As for the thumb, that is simply a synecdoche for human power. But it is not merely a figure of speech; according to the anthropologists, it was the opposed thumb that gave us an edge over the apes and supplied the basis for civilization. The thumb is the first tool by which we alter nature to suit ourselves. The gardener is reminded of this constantly, because the thumb plays such a prominent part in most gardening operations. He sows with his thumb and his index finger, doling out seeds by rubbing one against the other. With his thumbs working in concert, he tamps the ground around his transplants, assuring the purchase of their roots on the earth. All season long he is squeezing thumb and index finger together, pinching off shoots and leaves, deadheading flowers, squashing beetles, picking fruits. And what are the gardener's ever-present secateurs but a mechanical amplification of the opposable thumb? The thumb is our chief tool in the garden, and if we wield it particularly well, such that nature's ways and our desires are made to rhyme, it may be said to be green.

A good time to observe the green thumb's ease in the company of nature is at planting. Watch the way he handles seedlings. Compared to the novice, who treats his young plants gingerly, alarmed at their tenderness, the experienced gardener seems almost rough with them. The novice will baby a seedling, lay it gently into the earth at the same depth it stood in its pot, and tenderly tamp the ground around its roots. The green thumb is liable to flip the pot upside down till the transplant drops out and then jam it hip-deep in the dirt. What sort of sympathy is this? Now watch him tamp the soil down with his thumbs—*hard.* Then he attacks what few leaves and shoots he hasn't already buried, snipping off all but a handful. When he's through, his seedlings look much the worse for wear, at least compared to the novice's, which look exactly as they did at the nursery. But come back the next morning and the beginner's transplants will be hanging limply down, a row of tottering pensioners too weak to support their own weight. The green thumb's transplants, meanwhile, will look bright and perky as schoolchildren.

In fact the experienced gardener's roughness with his plants *is* a form of sympathy. He understands their predicament—that their roots have been so discombobulated they can no longer draw enough water from the soil to replace the moisture evaporating from their leaves. He knows that no matter how well he waters them, the transplants will die of thirst unless he amputates most of the leaves through which the water is escaping. The reason he tamps the soil so hard around the plant is to force root and earth into contact, and so speed the process by which the roots will put out new filaments and begin to draw water again.

Yes, there are books that will tell you all this—that you need to restore a reasonable "root/shoot ratio" if your seedlings are to withstand the shock of transplant. But the green thumb has an intuitive sense of these matters—of just how much shoot and leaf to amputate, and of when more radical measures may be called for. It is not too much to say he suffers along with the freshly planted tree, watching its leaves go limp and fold along their midribs as it struggles to staunch the flight of irreplaceable water molecules. When the dry west wind blows he can all but see those molecules lift from the leaves, and he knows the roots at that moment are as useless as fish gills in open air. He comes to the tree's rescue not with a hose but with pruning shears and saws.

Sympathy allows the green thumb to cut the tree's limbs back hard—that, and a sense of where a plant's "being" resides that is very different, I think, from the novice gardener's. The inexperienced gardener is loath to chop away at his new tree (indeed, to prune in general) because he assumes that the tree is indistinguishable from its limbs. This is probably because he looks at a plant more or less anthropomorphically—by means of a model that, though useful in some respects, fails to take account of those parts of the plant that don't, like men, stand on the surface. Later, if he gardens attentively and sympathetically, he'll develop a more complicated and less anthropomorphic understanding of how, and where, a tree lives. He'll probably come to think of the tree as having something akin to a soul that is distinct from its parts, and for which the limbs sometimes (at transplanting, for instance) represent a burden it may be glad to be rid of. If "soul" seems too mystical a term, think of it as simply the tree's life force, or the wellspring of its growth, what Dylan Thomas called its "green fuse." Imagine this as the fulcrum of the tree's roots and its visible parts, something located maybe just below ground level, and pruning will no longer seem cruel but beneficent, a form of relief and a spur to fresh growth.

Now, I have no idea if there is any "scientific" basis for these notions of plant identity, but I can't say I really care. It is enough to know that since I have begun to imagine my plants in this way I have had more success with them. The successful gardener, I've found, approaches science and folk wisdom, even magic, with like amounts of skepticism and curiosity. If it works, then it's "true." Good gardeners tend to be flat-out pragmatists not particularly impressed with science.

The fact is, science has done little to earn the gardener's respect. The miracle pesticides it gave him turned out to be a curse. Its picture of the garden's workings, and of the relationships between plants and soil in particular, turned out to be sketchy and incomplete. Science replaced the gardener's traditional picture of his soil as something alive and incomparably complicated—a mystery of fertility beyond human comprehension—with a chemical model that we now know is far too reductive. The fertility of a soil, science said, is simply a matter of its nitrogen, phosphorus, and potassium content; whatever was lacking, fertilizers could supply. But the scientist's truth turns out to have been, if not false, then dangerously partial: these elements *by themselves* did not produce healthy crops; maybe they tell 95 percent of the story, but not all of it. As the gardener always suspected, the soil *is* a mystery, a complex biological (and not just chemical) wilderness that we can nourish but not, as the scientist's black-and-white picture of it would have us think, simulate. The green thumb wants a more richly colored and nuanced portrait of his soil, one that takes account of the other 5 percent—that indicates how *alive* his soil is. And so in spring you'll catch him bringing a handful of earth to his nose, perhaps even tasting it, and then rubbing it in his hands to see what color it dries. The evidence of his five senses tells him things about the soil's condition and fertility that no laboratory ever could.

The Wall Street Journal not long ago published an amusing feature story about biodynamic farming, a school of organic agriculture given to various mystical theories and practices—planting by the phases of the moon, curing compost in stag bladders, harnessing various cosmic forces to improve crop yields. As you would expect, the article took the form of an extended smirk. I'd previously read around in biodynamic literature, even ventured one of Rudolph Steiner's murky theosophical texts, and had found much to smirk at myself. But some of the farmers quoted in the article made a certain loopy sense. If the phases of the moon can influence the physics of the sea, and even the biology of women, then why not also the growth of plants? "How can a rose be so perfect?" one farmer asked. "I wasn't satisfied with the material explanations of the world." It was enough for him that his biodynamic preparations seemed to work. And as the reporter had to admit, this man's crops—which likely had been planted at the full moon and nourished with a compost that had spent the solstice in an animal skull buried by a stream—were impressive indeed. Most biodynamic farms, in fact, produce uncommonly productive and healthy crops without the use of chemical fertilizers or insecticides. The farmers interviewed in the article reminded me of Woody Allen's story about the man whose brother thought he was a chicken. When a psychiatrist suggested he have his brother committed, the man shrugged, "I would, but we need the eggs."

No green thumb would dismiss biodynamic theories out of hand. Before he ridiculed a recipe for compost made from dandelions aged in a cow's membrane, he'd probably ask to taste a carrot grown in the stuff. If something works, that's all he needs to know—it matters to him not at all whether the idea came from Rudolph Steiner, the county agricultural agent, a Nobel laureate, or the old lady up the road who claims that the secret of her great asparagus is heavy applications of road salt. Actually, he's apt to try her idea first—she lives nearby (the best gardening knowledge is local) and, besides, he's seen the eggs.

Another way to define the green thumb might be as a gardener at ease amid science's lacunas. The mystery we impute to him is but an index of all that we don't yet understand. Compared to most of us, the green thumb is a patient and respectful student of the unaccountable 5 percent. He knows that in fact there *is* no material explanation for the beauty of the rose. If sometimes he lends an ear to mystical-sounding garden lore, that's only because these tips and stories reflect the success of many other gardeners over many years. In the garden, the voice of experience—distilled, collective, and well worn—sometimes speaks to us in the idiom of old wives.

It's the abstract, heroic voice of theory that the green thumb distrusts. He knows that to lean heavily on abstraction in the garden is just another form of overcultivation. Theories, whether they be scientific or astrological, are made by people and imposed on nature—they're all thumb. And though a theory may be powerful and contain truth, it is never more than a representation of nature. The good gardener knows better than to mistake it for the real, green, irreducible thing.

Observe the green thumb at work for a while and you'll notice how, in keeping with his preference for experience over abstraction, he approaches nature more like an artist than a scientist or engineer. He welcomes in his garden not only the laws of nature, but the play of contingency, too. He's open to happy accidents, more comfortable with cases than axioms, less inclined to analysis than to trial and error. Confronted with a problem—what shall he plant under the *Clematis jackmanii*?—he tries this or that, sees what happens, then tries something else.

In this the green thumb's creative method mimics nature's own. Within the world of his garden he plays the role of natural selection, except that the standard of fitness he applies is shaped by cultural as well as natural considerations—by his taste, for example. Whether the Queen Elizabeth rose survives in his garden has less to do with its hardiness than with his judgment of its color and temperament in combination with others. He tries the Queen Elizabeth with the *jackmanii,* and finds the combination jangling—the well-heeled grandma-pink of the rose seems almost to be waging class war against the clematis's forthright farmhouse pigment. So the next year he rips out the rose; he might try it elsewhere in the garden, but probably he throws it out: she's just too much of a dowager ever to be happy in this place. One winter he happens across an attractive, plain-spoken lily in the catalogs, Golden Splendor, whose clear, flat yellow looks like a perfect match for the *jackmanii.* That July he's pleased to find that the two plants' colors and temperaments do indeed rhyme, their bloom times coincide, and so the combination becomes a fixture of his garden.

When I say his method resembles evolution's, I don't just mean he practices a form of survival of the fittest. Evolution has a double rhythm: only *after* nature, in her promiscuous creativity, throws up countless new possibilities and combinations does natural selection (her critical impulse, if you will) step in to determine which work best under the circumstances. This colossally wasteful and extravagant process is what makes possible the extravagant beauty of the rose, which far exceeds any possible utility—much more modest flowers attract bumblebees equally well. Nature creates without an end in view; fitness is but an afterthought. The gardener in his own little world, like the artist in his, performs both functions, hatching the trials and then culling the errors.

But as much as he seems like a god in his garden, practicing his own local brand of natural selection, the green thumb entertains no illusions of omniscience or omnipotence. If he's any kind of god it's a Greek god, one whose power is sharply circumscribed by the willfulness of men and other gods. Unlike Yahweh, Athena bargains, cajoles, even loses one now and then; mortals can keep secrets from her. The green thumb knows he doesn't pull all the strings in his garden and, equally important, he prefers it that way. Indeed, he suspects that the garden over which he exerted absolute mastery would be a pallid, thin, uninteresting place. For isn't it precisely the partialness of our mastery—the unstillable rub of nature against culture, of fact against idea—that gives gardens their savor?

You might say that the green thumb gardens with a greater-than-ordinary measure of what Keats, seeking to account for Shakespeare's genius, called negative capability—the ability to exist among "uncertainties, mysteries, doubts, without any irritable reaching after fact and reason." The green thumb is equable in the face of nature's uncertainties; he moves among her mysteries without feeling the need for control or explanations or once-and-for-all solutions. To garden well is to be happy amid the babble of the objective world, untroubled by its refusal to be reduced by our ideas of it, its indomitable rankness. Can there be much doubt that Shakespeare had a green thumb?

But though the green thumb's approach to nature resembles the artist's in some ways, it doesn't necessarily follow that he thinks of his garden as a work of art. He's apt to be suspicious of the term, particularly of the closure or completeness it implies. (Good gardeners do not necessarily make good garden designers, and vice versa.) The green thumb accepts that a garden is never finished, that though he may tame nature for a time, his mastery is temporary at best. The manicured hedge will soon be rank, the garden path overgrown with weeds; rain will topple all the irises. What sort of artwork is this that so stubbornly refuses to keep still? If he is an artist, then he's like a sculptor whose stone responds to the strike of the chisel by growing a new appendage, or melting. Indeed, it is a measure of the gardener's negative capability that he can go on gardening in the face of such intransigence on the part of his materials.

More than a work of art, I like to think of the garden as if it were a capitalist economy, inherently unstable, prone to cycles of boom and bust. Even the most prosperous times contain the seeds of future disaster. A flush year in the perennial border usually means lean times ahead; now spent, the perennials need dividing and won't peak again for another two years. Unless pruned in spring, my asters, phlox, and delphinium will put out way too many shoots, a form of herbaceous inflation that will cheapen all their blooms come summer. Wealth is constantly being created and destroyed in the garden, but the accounts never balance for very long—a shortage of nutrients develops in this sector, a surplus in that one, the value of water fluctuates wildly. Who could hope to orchestrate, much less master, so boisterous an assembly of the self-interested? The gardener's lot is to try to get what he wants from his plants while they go heedlessly about getting what they want. At the risk of straining the metaphor, think of the gardener as something like the chairman of the Federal Reserve—powerful certainly, but far from omnipotent. The best he can hope to do is smooth out the peaks and valleys of his garden's cycles, restrain the lythrum's rampant growth, stimulate a depressed campanula, channel the territorial greed of his artemisia 'Silver King'.

The garden is an unhappy place for the perfectionist. Too much stands beyond our control here, and the only thing we can absolutely count on is eventual catastrophe. Success in the garden is the moment in time, that week in June when the perennials unanimously bloom and the border jells, or those clarion days in September when the reds riot in the tomato patch—just before the black frost hits. It's easy to get discouraged, unless, like the green thumb, you are happier to garden in time than in space; unless, that is, your heart is in the verb. For the garden is never done—the weeds you pull today will return tomorrow, a new generation of aphids will step forward to avenge the ones you've slain, and everything you plant—everything—sooner or later will die. Among the many, many things the green thumb knows is the consolation of the compost pile, where nature, ever obliging, redeems this season's deaths and disasters in the fresh promise of next spring.

Am I this gardener? Not yet, not yet. I still careen from blunders of undercultivation to blunders of overcultivation. What green thumb would ever, out of some misguided liberal notion, offer to share his annual bed with weeds, or let a woodchuck drive him to the point of firebombing his burrow? I remain timid with the pruning shears, too quick to reach for the sprayer, and I find myself yearning for a day when my garden will be finished once and for all. The refusal of this land to conform to my ideas for it—even just to *sit still* for a while—frequently drives me crazy. It would be fair to say that, unlike Shakespeare, I am given to "irritable reaching after fact and reason." Irritable, indeed.

And yet I know there have been moments when I've seen my garden as the green thumb would, times when I've moved among my plants with his ease. There is a mood that sometimes overtakes you in the garden, a form of consciousness, even, that feels like nothing so much as a waking dream. I imagine most gardeners have known it at one time or another. Maybe it is a late afternoon in July, and you have been busy in the garden at a variety of small tasks—snipping the spent blooms off day lilies, pulling weeds, pinching suckers from the tomatoes just now setting fruit, cutting back a leggy nepeta to promote a second bloom. You are working intently, and though sweat may be beading your brow, the work feels as effortless as puttering. Your tools are light in your fingers and your fingers know what to do. This delphinium needs a few shoots snipped off to concentrate its bloom; that clematis wants to be shown where to twine its vine. As your hands work the world retreats. It is exactly as Marvell has it in "The Garden," the mind "Annihilating all that's made/ To a green thought in a green shade."

Green thoughts, green fingers. The gardener's sublime, you might say, and a very different one it is from the romantic's version, awestruck in nature and disarmed, or Zen's hollowed-out egolessness. The gardener doesn't lose himself, much less his body, in his particular reverie. For as much as you are being acted upon by everything around you this July afternoon, you are acting too, urging as well as listening, conducting that desultory conversation with nature that is gardening in summertime. And though it may not last very long (watch me fall into argument or declamation), at times like these the green thumb's way through nature seems clearly marked, easy to follow, almost second nature. So simple: grace in the garden but a form of puttering.

Fall

CHAPTER 8

The Harvest

With the harvest moon, which usually arrives toward the end of September, the garden steps over into that sweet, melancholy season when ripe abundance mingles with auguries of the end anyone can read. Except, perhaps, some of the tropical annuals, which seem to bloom only more madly the closer frost comes. Mindless of winter's approach and the protocols of dormancy, the dahlia and marigold, the tomato and basil, make no provision for frost, which might be a month away, or a day. The annuals in September practice none of the inward turning of the hardy perennials, which you can see slowing down, taking no chances, turning their attention from blossom and leaf to root and stashed starch. But instead of battening down the hatches, saving something for another day, the annuals throw themselves at the thinning sun, open-armed and ingenuous. On those early autumn days when frost hangs in the air like a sword of Damocles, evident as sunlight to the lowest creature, is there anything much more poignant than a dahlia's blithe, foolhardy bloom? When the mildest frost, one of those tentative breaths of winter September often brings, could blast it black overnight?

The harvest moon sometimes ushers in such a frost, always one of nature's heartbreakers, since typically it is followed by a few weeks of fine growing weather. When the tomatoes have succumbed to a September frost, and hang like black crêpe from their cages, those weeks can seem cruel—the tease and rebuke of missed opportunities. So on those evenings when a full moon dominates a cloudless sky, and the air has a faint metallic tang to it, implying it will give up its heat without a struggle, we make a last-ditch stand on behalf of the annuals. To hold close some remnant of the earth's warmth, we dress the tomatoes and squashes and cucumbers in old bedsheets and tarps. On silvery nights like these the vegetable garden looks like a congregation of ghosts, and the earth feels like it's lost its blanket; nothing stands between it and outer space. Bedsheets, a tender annual's spacesuit.

With luck, the garden slips past these few chilly nights into a string of safe, warm days. In the season's slanting light, the whole garden looks overripe, laden, and slightly awry. The sunflowers have blossomed massively, and now nod, drowsier by the day, their heads too heavy to hold up to the sun. Jays perch on their rims, hanging upside down in order to peck off the fat seeds. The presiding color of the season is a sharp orangey yellow—the acid shade of squash flesh, sunflower and black-eyed Susan petals, sugar maples' turned leaves. Of Mongol pencils and school buses, too, for isn't this the official hue of all things back-to-school?

No time now for summer's idle puttering, there's real work to be done in the garden. Harvesting is the least of it, if still the best. Now's also the time to dig new beds, plant trees and shrubs, spread compost, rake leaves, plant cover crops. Summer's work fingers and secateurs can handle; autumn's wants spades and forks, the commitment of arms and backs. And the weather obliges, with cool, brittle days on which it is a pleasure to sweat.

I do what I can, but I have to admit that as autumn advances and the sun's arc slips farther and farther south, my heart is not always in it. At least not in the big, new, forward-looking projects that the garden columnists in the newspapers, and the sales at the nurseries, exhort us now to undertake. Every year I'm reminded of the wonderful things I can plant in the border to extend the season of bloom. Sounds to me like denial . . . and maybe a bit of greed, too. Business is slow in the nurseries this time of year, so the garden world tries to persuade us that our ancient instinct is mistaken—that *fall* is really the time to plant. Trees maybe, but in this part of the country the gardener who plants perennials in October will be planting them again in May—because they won't have had enough time to settle in before winter mounts its assault. But it's not only that. As the earth prepares to close up shop, I'm just not in the mood to argue with nature's agenda.

Sir James G. Frazer, in *The Golden Bough,* tells of a North American tribe, the Esquimaux, that holds a contest each fall between the forces of summer and winter to see which will prevail. The tribe divides itself into two parties, the ptarmigans and the ducks; the ptarmigans are those members of the tribe born in winter, the ducks those born in summer. Each team grabs hold of one end of a sealskin rope and a colossal tug-of-war ensues; if the ptarmigans win, winter will come. You can't help thinking a victory by the ducks rings a little hollow, that their whoops might lack a certain conviction. Avid fall gardeners remind me of the Esquimaux's duck party. For a while, through September, say, I can see pulling against the forces of winter, dressing my tomatoes in spacesuits to buy them a little time. But at a certain point every fall I feel like letting go of my end, throwing in with the ptarmigans.

Unless you are a duck or a tropical annual unversed in northern ways, nature's agenda in the fall garden is hard to miss: prepare for winter. As the supply of available light dwindles, the work of photosynthesis winds down and the plants concentrate their remaining energies on ripening fruit and seeds. To the gardener, the green world suddenly seems much more manageable. The grass is slow to replenish its ranks after a mowing, pulled weeds no longer reappear overnight, and the gardener at last can get out ahead of the green parade and not have his works trampled.

In the garden and the woods the color green loses its overwhelming majority for the first time since May. With ripeness comes an outbreak of individuality, as each fruit acquires its own signature tint. The tomato vines shrivel to wan strings while their fruits swell red and yellow; the orange shoulders of carrots shove up from the ground, and the various shades of winter squash—spaghetti's yellow, tan butternuts, hubbards the color of a robin's egg—show forth as their tangle of foliage collapses around them like discarded clothes.

Autumn color in the woods signals the abdication of chlorophyll; in the garden, among the annuals, it means something else. With their ripe, tinted fruits the plants aim to flag down passing animals, offering them food in exchange for giving their seeds a lift. By late September the plants are concentrating all their energies on this process—on writing down their secrets on tiny seed tablets and then encouraging someone, *anyone,* to take them out into the world. Recipes, instruction manuals, last testaments: by making seeds the plant condenses itself, or at least everything it knows, into a form compact and durable enough to survive winter, a tightly sealed bottle of genetic memory dropped onto the ocean of the future. (In the case of those trees which form nuts, the survival of genetic memory depends on forgetfulness; if the squirrels that collect and hoard acorns did not forget where they buried so many of them—50 percent, in Beatrix Potter's estimate—oaks would be no more.)

As for passing animals, there's no shortage of these. The scents and hues of ripeness in the garden set off a scramble for its fruits—preparation for winter being the animals' agenda as well. The woodchucks and raccoons, deer, squirrels and moles rouse from their summer lethargy and pitch themselves into one last great battle for the season's spoils. Carbohydrates, fats, and proteins by all rights mine they consider up for grabs, and their autumn assaults can rival frost's for destructiveness. The last year I planted corn, I hadn't harvested more than a half dozen ears before a gang of raccoons climbed the fence one night and threw a raucous party on my tab. They toppled every single cornstalk, ruining the crop yet not even eating it all—half-chewed ears littered the garden like empties. It looked as though they'd take a bite or two from an ear, fling it over their shoulder, and then reach for another. They stomped through the beds, ripped the tops off the leeks and beets strictly for spite, and then deposited several turds—large, *impudent* turds—smack in the middle of my beds.

Compared to the cat burglaries of deer and woodchucks, this looked like the work of the Manson family.

But the raccoon's debauch is only an extreme and wasteful instance of a widespread practice at this season of harvest high and low. For at the same time I'm kept busy fending off the mammals, there are also fungi and bacteria to contend with. Ripeness beckons them too and, under the broad rubric of rot, they probably claim more of the harvest than the mammals and I combined. As epidemiologists well know, at this point in our evolution the microorganisms offer our species its only significant competition; compared to the bacteria, fungi, and viruses, the raccoons and woodchucks, the lions and tigers and bears, are a joke. It's the ones you need a microscope to see that can really mock our claims to having mastered nature, or "conquered" disease.

In the garden the microorganisms wrest ripe fruits from us by rendering them unpalatable or poisonous, thereby ensuring that we'll leave them to rot on the ground. The soft, pussy beachhead that bacteria stake out on a tomato; the neat, slowly expanding target fungus draws on squash; and the black spots a virus stamps on apples—all these are flags of victory hoisted by microbes that go uncontested by us. In the description of one biologist, the microorganisms operate like the child with a plate of cookies set before him, making sure to lick every one so no one else will want them. Even the animals are repelled, particularly by the scent of fermentation, possibly because evolution has dealt harshly with animals who liked to get tipsy on alcoholic fruit.

Considered from this aspect, the autumn garden holds many horrors, particularly in a damp year. The tomato vine that, laden and bending low, touches a ripe fruit to the ground also dooms it; a day will turn it watery and repulsive. Reach your fingers around the base of an overripe cabbage and you might close them on a mass of slippery brown pus that, unseen, has been eating into the head from below. They Came from Beneath the Earth, it seems (though the airborne spores of fungi are winging in from all directions); the earth fairly bubbles with rot this time of year, a putrefactory reminding the gardener that whatever it gives it can also take back. The ground is littered with soft, slowly darkening lumps of decomposing vegetable matter that each day grow less distinct, their form and color gradually fading until they are dissolved back into the soil entirely.

Walt Whitman marveled that we should ever dare to touch the earth, much less ever remove our clothes in the presence of its corruptions. "How can it be that the ground itself does not sicken?" he asked in a poem called "This Compost":

> *Are they not continually putting distemper'd corpses within*
> *you?*
> *Is not every continent work'd over and over with sour dead?. . .*
> *What chemistry! . . .*
> *That when I recline on the grass I do not catch any disease,*
> *Though probably every spear of grass rises out of what was*
> *once a catching disease.*
> *Now I am terrified at the Earth, it is that calm and patient,*
> *It grows such sweet things out of such corruptions. . . .*

Harvest's work is to hold off, at least temporarily, earth's corruptions, the spoilage of our spoils. So we pick all the fruit we can before the animals get to it, and then deploy a variety of ingenious techniques to thwart the microbes. Cooking, canning, freezing, acidifying, smoking, salting, sugaring—the culture's time-tested prophylactics against nature's rot, ingenious tools of the "kitchen garden." Some of our most satisfactory methods of preservation, in fact, work on the same principles as gardening. In making wine or hard cider or various kinds of cheese, we don't so much battle nature's microbes as pick and choose among them and then let them work to our benefit. We harness the processes of decay, garden rot itself.

Nature's gothic—but of course that's only part of the season's story. Harvest isn't all scramble and rot, spoils and spoilage. There is the fact of abundance, too, the season's freely given, unencumbered gifts—"teeming autumn, big with rich increase" (Shakespeare, in Sonnet 97). On one of those cold and extra-vivid October afternoons, when a hard frost threatens and we're hurrying to harvest all but the hardiest vegetables, autumn's big increase never fails to astound. So much sheer, indubitable *mass,* none of which existed just a few months ago except in the prospect of a handful of seeds. We heap bushel baskets with summer squash, cucumbers, tomatoes; stuff bags with lettuce and chard; cut whole heads of sunflowers big as a calf's, and haul it all indoors, where it commandeers the kitchen. But beyond the impressive bulk, there's the unexpected *weight* of it all—almost as if we're shouldering not just baskets of produce but fall's very gravity itself, the same ripe force that bows the sunflower heads and bends low the boughs of the apple trees that ring my garden. Apples especially seem vested with the season's extravagant gravity. Pliny said that apples were the heaviest of all things, according to Thoreau, and that oxen start to sweat at the mere sight of a cartload of them.

They might very well sweat, too, at the sight of one of the winter squashes I discovered in my garden this October. I didn't fully appreciate the magnitude of it until then, after the mass of foliage beneath which it spent the summer fattening itself had shriveled. Without a doubt the biggest thing I've ever grown, this squash tipped the scales at close to thirty pounds. Its seed I had obtained last winter from a firm in Idaho that specializes in "heirloom" vegetables—old varieties no longer grown commercially. Called 'Sibley', my squash is said to be an American Indian cultivar that was passed on to the early settlers. The reason it fell from commercial grace, I'd guess now that I've laid eyes on one, probably has to do with its looks, which are definitely on the homely side. A Sibley is a big warty thing with the washed out, blue-green color of dirty ice; it might be a chunk of glacier. Its form, however, is pleasing, sort of: pinched at both ends and bulging at the waist, it looks like a gondola, or a Viking ship, listing under a fat cargo. Or like a crescent moon with the belly of a Buddha.

Where did this thing, this great quantity of squash flesh, come from?

The earth, we say, but not *really;* there's no less earth here now than there was in May when I planted it; none's been used up in its making. By all rights creating something this fat should require so great an expense of matter that you'd expect to find Sibley squashes perched on the lips of fresh craters. That they're not, it seems to me, should be counted something of a miracle.

The first person to verify that indeed this is a miracle was a seventeenth-century Flemish scientist by the name of Van Helmont. He planted a willow sapling in a container that held 200 pounds of soil and, for five years, gave it nothing but water. At the end of that time, the tree was found to weigh 169 pounds, and the soil 199 pounds, 14 ounces—from just two ounces of soil had come 169 pounds of tree. Rich increase indeed.

Before I harvested my Sibley and stopped to consider its provenance (and read about Van Helmont's experiment), I had always thought of gardening as a zero-sum enterprise—that it was necessary to add as much to the garden (in the form of nutrients) as the produce I harvested removed from it. I assumed that I'd have to replace whatever my giant squash took from the soil or eventually nothing would grow in it. And though it is true that a monster squash like mine does deplete the soil of certain elements, their quantity is negligible; a small handful of compost could easily cover the deficit. But that deficit is much smaller than the sum total of matter my squash represents. Were I to leave it to rot on the vine, there would actually be a surplus in the garden's accounts; the soil would be both richer in nutrients and greater in total mass than it was before I planted it. Much of the increase is water, of course. But the remarkable fact is, my Sibley, considered from the vantage of the entire planet's economy of matter, represents a net gain. It is, in other words, a gift.

This is not exactly news, I know; Van Helmont could have told you as much three hundred years ago, Shakespeare evidently sensed it, and so did all those Renaissance painters of cornucopia. But it's something we seem to have forgotten in recent years, as our concerns about the depletion of the Earth's resources have mounted. We take it as an article of faith today that the Earth is running down, that we are using up its finite supplies of energy, fertility, and resources of all kinds. We've come to think of the Earth as a closed system; one of the age's presiding metaphors is "spaceship Earth." Conceived as such, it's easy to imagine the ship's provisions gradually being exhausted; as more and more matter is converted into energy, we must eventually run out.

Entropy is the great faith of our time. Those who are most awed by it preach "limits to growth"—that we should consume our fixed, unreplenishable stores as slowly as possible. On a spaceship, this makes good sense. But the second law of thermodynamics, under which entropy increases as matter converts to energy, applies only to closed systems, and, as the environmentalist Barry Commoner points out, the global ecosystem is not a closed system. The Earth in fact is nothing like a spaceship, because new energy is continually pouring down on it, in the form of sunlight—free, boundless, virtually infinite sunlight. And sunlight come down to Earth is used by the process of photosynthesis to create new plant matter. Plants, in other words, are energy returned to matter—entropy undone, at least here on Earth.

The lesson in this is not that we should feel free to waste our resources; it's that our environmental problems may have more to do with our technologies and habits and economic arrangements than with the planet's inherent limits or the burden of our numbers. All we could ever possibly need is given. In terms of the global ecosystem, there is a free lunch and its name is photosynthesis. In a sense, the ancients were entirely correct to regard the harvest's abundance as a gift from the heavens; and I would not be too far off regarding my squash's lunar silhouette as a reminder of its extraterrestrial origins.

Living as we do in the autumn of a millennium, and feeling somehow that we've come very late into this world, this strikes me as the harvest's most salutary teaching—indeed, as reason enough to garden. Here in my garden the second law of thermodynamics is repealed. Here there is more every year, not less. Here it is ever early, never late. Here, in the ungainly form of a Sibley squash, newness comes into the world.

By late October, after I've harvested as much as I can reasonably expect from the garden, killing frosts have ceased to terrify. The deaths of the annuals are stingless now, almost a release. On the morning of the first really hard frost the vegetable garden is more black than green, and the whole edifice, from the broad tenting leaves of the squashes to the upright networks of bean and tomato vines, sags as though all the air had suddenly been let out of it. It's not really air, of course, but water that's been holding up the garden since spring. Except for the woody plants in them, gardens are just elaborate architectures of water; cells tensed with liquid and stacked like bricks give it its form. Frost blasts that structure cell by cell, as the glass-sharp facets of the ice crystals it forms within puncture each cell wall, releasing the water trapped inside. Everything, you realize, depends on this precarious tumescence. Though the cells don't actually begin to lose their form until the temperature rises again; if it stayed below freezing, the plants would remain stiff and green indefinitely. It's the warming rays of the morning sun that betray the wounds within, releasing summer's taut waters and collapsing the green bodies into black heaps.

Some mornings now the ground is iron, other days still workable. The earth at this season is a gate that swings open and shut in the stiffening autumn wind until one very cold day, coming usually in December, it catches and stays closed. But it's well before the earth locks that I feel ready to go out from the garden. For me the flower garden loses its hold in October. There's still plenty of bloom then—the asters, heliopsis, and rudbeckias hold gamely on in the perennial border, and among the annuals, the snapdragons can be counted on to eke out a few blossoms right into November. But the late flowers just can't hold the eye once the forest has started to bloom.

As radical as frost, the trees of a New England autumn change everything in the garden. In the same way that serious cold seems to invoke a new set of physical laws—altering the way sound carries, making the air less elastic—the coloring of maples and hickories and oaks overturns laws of space and light that have been in force since spring. All summer, the matte green walls of trees that enclose my land lend the garden its intimacy, establishing a scale that is flattering to garden plants; the summer trees form a neutral stage on which the flowers can put on their genteel show without competition from the larger landscape. Now, though, the green room seems to grow larger every day, as each new scrim of forest color draws the eye farther out through what was formerly opaque. The domestic scale of outdoor space in summer gives way to something grander. The walls lose their modesty, flower spectacularly, and the perennial border all but fades away, like an unlucky bride who finds herself overshadowed by her bridesmaids' more brilliant gowns. Of the fall foliage in New England, Thoreau said that "if such a phenomenon occurred but once, it would be handed down by tradition to posterity, and get into the mythology at last."

For the whole forest now is a flower garden, the hickories its lemon lilies, the sugar maples its blushing dahlias, the scarlet oak its rose. Quickest to fall, the ash leaves freckle the lawn like daffodils in April; a few weeks later, the Norway maples drop bright yellow skirts around their ankles. Now when the sun is low in the sky, its slant rays snag in the tops of the scarlet trees, setting forest fires from west to east, a border of blazing crowns miles long. By mid-October, the walls of my garden are in furious revolt; the common species of the forest have run roughshod over the flower beds, torching the precious hybrids, and wresting the attention of our eyes.

Against this, I won't even try to defend the garden. I'm inclined for once to agree with Thoreau, who argues in a fine, little-read essay called "Autumnal Tints" that "Comparatively, our gardening is on a petty scale,—the gardener still nursing a few asters amid dead weeds, ignorant of the gigantic asters and roses which, as it were, overshadow him, and ask for none of his care. . . . Why not take more elevated and broader views, walk in the great garden; not skulk in a little 'debauched' nook of it? consider the beauty of the forest and not merely of a few impounded herbs?" Here, amid the easy extravagance of the October forest, the perennial border does seem slightly absurd, spewing forth its color like a fountain in the middle of the sea.

Thoreau wrote relatively little about autumn until this essay, composed in the last months of his life. Go back to *Walden,* which purports to be a chronicle of an entire year, and you won't find more than a few paragraphs about autumn; that's because, metaphorically at least, *Walden* is a book about springtime. About renewal, fresh prospects, and defiance, too, and autumn tends to undermine exhortations on such themes. Too much talk of fall in *Walden* would have collapsed that book's taut, unflagging spirit like a hard frost. Thoreau, like his mentor Emerson, for the most part kept his moments of resignation confined to his journals. At least until those last months when, dying of tuberculosis, he took up the subject of autumn leaves. "How beautifully they go to their graves," he writes. "How gently they lay themselves down and turn to mold. . . . They teach us how to die. One wonders if the time will ever come when men, with their boasted faith in immortality, will lie down as gracefully and as ripe,—with such an Indian summer serenity will shed their bodies, as they do their hair and nails."

Autumn's no season for defiance. You can dispute nature's agenda all you want, play tug-of-war till you're blue in the face, but the duck party never wins, not really. So if I feel like giving up now, if I feel like shedding for a time the cares of this garden and following Thoreau out into that wider one, the October forest, I will. To do so is not to forsake my garden, only to acknowledge the temporariness of my hold on it, and the inevitability of its demise. In "Autumnal Tints" Thoreau is not making his usual case for the moral superiority of wilderness; he mentions that the maples on the Concord green move him as much as those in the forest. No, his real subject in that essay is fate.

A garden that never died eventually would weary; maybe gardens require walls in time as well as space. The garden winter doesn't visit is a dull place, robbed of springtime, unacquainted with the extraordinary perfume that rises from the soil after it's had its rest. That promise, the return every spring of earth's first freshness, would never be kept if not for the frosts and rot and ripe deaths of fall. I don't think I want to stand in the way of this. (As if I could!) So when I go out from the garden for the last time in autumn, I leave the gate open behind me.

CHAPTER 9

Planting a Tree

For some time now, I've been thinking about planting a tree here—a *real* tree. It's not that I haven't planted any trees before (I've probably planted two dozen), but all of these have been minor trees, lightweights really, the kind you can justify in the short term: white pines to screen the road, dwarf fruit trees, a crab apple or two, a tree hydrangea, and a pair of *Salix babylonica,* the Tree of Immediate Gratification, otherwise known as the weeping willow. For a long time, this was my idea of a tree: something you could pick up for $29.99 at the nursery, stuff in the rear of a hatchback, jam into any old hole, and then virtually watch the thing grow. It took less than three years for my hatchback-sized willows to blow themselves up to the size of hot-air balloons.

Not to take anything away from my willows, but they do lack a certain . . . *gravitas.* And there was little else on the property to supply that element. The biggest trees on this former farm are the pair of white ash that stand at the head of the driveway. Although a good fifty feet tall now, they're nevertheless peculiarly unobtrusive. Bigger than a town house, yet you'd hardly know they're there. But that's ash; their trunks can rise thirty feet before branching, and even when fully fledged, they pass along most of the sunlight that comes their way. Each year they get on and off the stage with a minimum of fuss, leafing out not until late in May and then shedding their foliage before the end of September.

Great trees, but self-effacing ones, happy to have other plants grow in their thin, dappled shade, and to supply people with excellent firewood, furniture, baseball bats, and tool handles—including handles for axes, which gives some idea just how accommodating ashes are.

When you think of a farm in New England, probably you picture a few venerable oaks or maples near the house. This was never that kind of farm, even though it's been under cultivation since colonial times. Beginning in the 1920s, and continuing up until just a few years ago, it was a dairy farm operated by a family that probably never extracted more than a tenuous, hand-to-mouth living from this land. A couple of sugar maples by the house would have told of a certain achievement of comfort in the Matyases' relationship to this land that I doubt they ever felt. It also would have told of an expectation of continuity here. But the children evidently had no interest in farming this pinched, craggy wedge of hillside, because when the elder Matyases died the farm was broken up and sold, piece by piece.

No one in town has a good word to say about old Mr. Matyas (the name is pronounced "Matches"); "so mean, he hated his self," is how one neighbor described him to me. Everyone will give him this, though: he made some of the best hard cider around, or the most potent, anyway. And in fact the only real tree planting he undertook was of a half-dozen apples, which today are the most beautiful trees on the property by far. Now more than a half century old, their crabbed and weathered forms suggest the character and long witness of architectural ruins. Some days they look like monuments to their planter's own legendary crabbiness. But obviously little aesthetic consideration went into their planting; the idea, apparently, was strictly utilitarian—to secure a free and reliable source of booze. The farmer left hundreds of gallon jugs in the root cellar.

No, this was definitely not one of those genteel New England farms whose owner had the leisure and forethought to plant oaks for the benefit of generations to come. When we bought the place there was something unmistakably Appalachian about it, the yard completely lacking in ornament (unless you count old tires and busted farm equipment) or any other evidence that its residents had taken pleasure in their land. Joe Matyas probably could not have enjoyed a patch of shade if he had one, and to provide one for his children would only have led to the sloughing of chores. Shade may have been a luxury the Matyas farm couldn't afford.

The absence of great trees underscores the harshness of the land, as well as the solitude of the little Sears, Roebuck house plunked down onto it. It was in part to soften this effect that I wanted to plant trees. I say "in part" because I'm beginning to realize that planting trees is a complicated act, born of many intertwined causes not easily teased out. But *one* of my motives was aesthetic. A great tree changes the look of the landscape, of course, and not only from a distance; it shapes space in the third dimension, too. An old sugar maple—that was the tree I had in mind—sponsors a distinct kind of light and air around itself. Its shade is dense, but always sweet, I think, and never oppressive. The space that a maple articulates seems particularly hospitable to people—it's an intimate, almost domestic space, compared to that of a venerable old oak, say, which will seem grander, more imposing. No matter how large it grows, a maple never drops its tie to the human scale; a few of its boughs invariably reach down to us so that we may climb up into them, if only in our imaginations. Maples suggest haven. They always look comfortable next to houses, in summer gathering the cool air close around them and then ushering it toward open windows.

A single great tree can make a kind of garden, an entirely new place on the land, and in my mind I was already visiting the place my maple made, resting in its shade. I knew it wouldn't happen overnight, probably not even in my lifetime, but wasn't that precisely the point? To embark on a project that would outlast me, to plant a tree whose crown would never shade me but my children or, more likely, the children of strangers? Tree planting is always a utopian enterprise, it seems to me, a wager on a future the planter doesn't necessarily expect to witness.

Just thinking about it in these terms was starting to make me feel rather virtuous, I have to admit. And as I drove to the nursery early one October morning, I began to form large conclusions about Our Age based on the fact that no one planted great trees any more. Who today can imagine a summer day in the next century, sitting in the shade of a maple planted in 1989? Not many, to judge by what we choose to plant these days. Gardeners in this country once planted trees with the kind of enthusiasm we bring to the planting of perennials today. What tree planting we do usually consists of marooning a few small ornamental specimens in a sea of lawn. True, we have less space to work in, and we move every seven years or so, but I can't help thinking some cultural pathology must be at work here, too, some failure of imagination as regards the future. (I've read that Germans stopped planting oaks and other slow-growing hardwoods in Weimar days.) Visiting the Huntington botanical garden in Southern California not long ago, I'd been deeply impressed to learn that the three magisterial cypresses that dominate the grounds had been planted from *seeds* that Henry Huntington had gathered in Chapultepec Park in Mexico City at the turn of the century. Now there was a confident age.

"To plant trees," Russell Page wrote in his memoir, "is to give body and life to one's dreams of a better world."

In the company of large and uplifting thoughts such as these, I went shopping for my tree. I told John, the nursery manager, that I was in the market for a shade tree, probably a sugar maple. John frowned and slowly shook his head, giving me one of those looks a car mechanic trots out after he's peered under your hood and found an afternoon's work. Apparently the sugar maples in the area had been having a hard time of it lately, and he strongly advised against planting one. It seems that the pear thrip, a microscopic insect, has infested sugar maples all over New England. The thrips feed on the buds in April so that the leaves that emerge in May are too small and gnarled to be much use. Badly infested trees are forced to produce a second set of leaves at a crippling expenditure of strength; after several years, the effort kills them. Thrips have always been around, John said, but the maples have become more susceptible lately, probably as a result of the stress acid rain puts on them.

This is just the sort of news that can suck the wind out of utopian sails. But now that John mentioned it, I recently had noticed several dead maples on local roads, monumental nineteenth-century plantings that last spring had simply failed to leaf out. It's depressing to think that smoke-stacks in Ohio might be the cause.

John told me I would be better off with a Norway maple, a European variety that thrived in cities and seemed to be relatively unperturbed by the stresses of civilization. He pointed out a few local specimens I was familiar with, massive oval crowns that flared bright yellow in autumn, and I decided a Norway maple would do.* John showed me his stock, a handful of fifteen-footers (each about two and a half inches in diameter) that sold for $129, another $10 for delivery. Even at that size, they were frankly not all that impressive; spindly poles, really, topped by a few forked twigs. Squaring my utopian picture with these glorified dowels wasn't going to be easy. Probably sensing my disappointment, John rested his hand on one of the trees at shoulder height and said: "But these Norways are some quick growers. Ten years, you could have a respectable little tree; twenty, maybe even see a bit of shade."

A "bit" of shade by 2010?! Suddenly I was beginning to feel discouraged about the whole enterprise. I felt less like Henry Huntington gazing confidently across the prow of the new century than Joe Matyas brooding on his cramped horizon and reaching for a drink. Maybe I'd be better off with an apple tree, or another willow . . . I mean, how long were we going to live in this house, anyway? But then I was recalled to my Large Thoughts—my high-minded desire to make a positive statement about the future. I swallowed hard and told John to deliver it tomorrow.

I spent that evening reading up on tree planting. All the books I consulted sought to impress upon me the enormous responsibility I was un-

* This was an even more fateful decision than I knew: The Norway maple, I later discovered, is an invasive species scorned by sophisticated American gardeners. It also casts a dense shade in which little will grow. The more beautiful native Sugar maple would eventually recover from its thrip troubles; even so, I'm standing by my unloved Norway, if only as an enduring badge of my horticultural naïveté.

dertaking, and I was duly impressed. Choosing the Site and Digging the Hole, in particular, were crucial, irrevocable acts that, handled badly, I would rue for decades.

It's a sobering responsibility, picking the site for a big tree; get it wrong, plant it too close to the house or an electrical line, and you will someday force a terrible decision on someone. To plant a big tree is to throw a long shadow across the future of a place, and we're obliged to consider its impact carefully. I spent half a day walking around the property, straining mentally to add something the size of a brownstone to the empty scene before me. (Could it be the fact that I will not live to see the mature tree that makes imagining it so hard?) I traced one fifty-foot circle after another in the grass, trying to picture the eventual footprint of shade. Shadows you can see are elusive enough; to plan for shadows decades hence is to deal in the shadows of shadows.

I settled on a spot in the middle of an open meadow about halfway between the house and the barn where I keep my office and Judith has her studio. This is a focal point on the property, visible from several rooms in the house, as well as from the barn and the driveway. Devoid of any shade or shadows, the meadow suffers a particularly harsh light and in summer feels hot and dry, inhospitable. We walk across it maybe a half-dozen times each day, and the prospect of a maple shading our path up to the barn is an attractive one. The tree will also command fine views both from the bedroom, where the slant rays of the morning sun will have to pick their way through its leaves coming in, and from my desk in the barn loft, where the late, reddening sun will light it up from behind. It's going to be something to see.

Early the next morning I began digging the hole, another solemn responsibility. In the same way that all future occupants of this house will have to live with the consequences of my choice of site, the quality of the hole I prepare will help determine the future well-being of my tree. Among plants, hole is fate. This fact was first seared onto my consciousness by Ralph Snodsmith, a sweet, somewhat anachronistic fellow from whom I once took a course (Gar 101) at the New York Botanical Garden.

Snodsmith, who drove a green Mercedes and wore a green suit and a green tie to every single class, imparted a few vivid rules of thumb that have proven hard to shake. "Xylem up, phloem down." "Keep your eye on the root/shoot ratio." And the one he repeated approximately every fifteen minutes and then tested us on in both the midterm *and* the final: "Better to put a fifty-cent plant in a five-dollar hole than a five-dollar plant in a fifty-cent hole."

Though there is some debate on this point, most of the books advised a hole as deep as the tree's root ball and twice as wide, which in my case meant a hole six feet wide and three feet deep. (In good soil, a smaller hole will suffice.) The line drawings in the book showed cross sections in which neat pyramids of soil stood next to equally neat inverted pyramids cut into the ground. They didn't show boulders the size of office safes. More rocks came out of my hole than soil, some so big I had to roll them out, pyramid-builder style, along inclined wood planks. More than once I thought about trying a different site, but finally decided that whatever glacier had strewn this igneous and sedimentary woe around the property had probably done so democratically. One of the earliest settlers in my town, after paying a discouraging first visit to the Cornwall land he intended to farm, composed a couplet in its honor: "Nature exhausted all her Store/to throw up Rocks, but did no more."

Digging in that ground—*mining* might be a more accurate word—left me a bit more sympathetic toward Joe Matyas and the farmers that preceded him here. If I depended on this land for my living, I might not have planted great trees either. Because you plant trees in ground you feel a certain affection for, and working this soil, so grudging of everything but rocks, was bound to elicit more bitterness than gratitude. Wendell Berry says that trees on a farm are a sign of the farmer's "long-term good intentions toward the place." Probably so, and my guess is that Joe Matyas's intentions toward this place verged on the spiteful.

Digging the hole took the better part of a day, but, taking frequent rests as I did, I had plenty of time to lean on my shovel and muse. Looking around at the society of trees my maple was about to join, I realized that they held a record of both the social and natural history of this land. The ancient, hell-bent apples not only memorialized Joe Matyas's tenure but their rings contained a chronicle of the last fifty years' weather; a child could probably pick out the "greenhouse" summer of 1988, a summer so hot and dry that the trees added their most slender ring of growth in a century.

Even in the short time I've been here, I've seen how each important event of natural history leaves its mark on my trees. One of the apples bears a nasty scar where a large bough was ripped from it during a freak blizzard on October 4, 1987, a notorious storm that proceeded from here to cross the Atlantic, attain hurricane force, and then, on October 16, cut down thousands of England's most treasured oaks and elms, great eighteenth-century plantings that were considered part of that country's patrimony. On July 10, 1989, the ash on the south side of my driveway had a forty-foot-long section of its bark unzippered by a bolt of lightning. Though it remains uncertain whether the tree will survive, I nevertheless count myself lucky: that same storm spun off a ferocious tornado that danced through Cornwall, uprooting or snapping the crowns off several thousand of the town's oldest trees. On July 9, Cornwall had a graceful, maple-lined Main Street that made it the picture of a nineteenth-century New England village; on July 11, the town, utterly denuded of trees, looked more like a frontier outpost that had been hacked out of the forest overnight. Cornwall won't look settled on the land again in my lifetime. Hoping to hasten that day, everyone in town is busy this fall planting trees (could this be another of my motives?), and my maple will join a large new generation of trees that will commemorate the 1989 disaster deep into the next century.

Like Joe Matyas's apple trees, my maple will also mark a turning point in the property's social history. I can't be entirely sure what my maple will signify to someone leaning on a shovel and musing fifty years hence, but I can guess: the arrival of a more cosmopolitan era on this farm, a time when its owner had the means and the leisure to plant a strictly ornamental tree. But what if I'm all wrong about this? What if planting a maple will mean something entirely different in fifty years, and my successor here will interpret the planting of this tree as . . . I don't know, as being quaintly arrogant, say, or "speciesist," because people by then had decided that trees have certain inalienable rights, one of these being the right not to be planted within fifty feet of any human habitation. Or maybe the oil will have run out by then, and the beauty of a stack of cordwood will far surpass that of a mature maple.

About now you're probably thinking I should have put the musing on hold and returned to my digging, but I'm afraid I didn't do that. Because my spade-side historicizing had made me begin to wonder if perhaps I'd been unfair to Joe Matyas. Could it be that for Joe chopping down a tree was fully as virtuous an act as planting one is for me? And that a truer history of this place would pay as much attention to the trees that *aren't* here as the ones that are? Perhaps it's not the apples, but the open meadows that constitute Joe's principal contribution to the moral economy of this place. As the owner of a marginal New England farm, Matyas's feelings about trees probably shouldn't be judged according to my (or Wendell Berry's) standard, but by the standard of earlier New England subsistence farmers, for whom a tree was at best a resource to be exploited and, at worst, an impediment to agriculture—a big weed. For most of the time that this land has been in white hands, cutting down trees has seemed as civilized a thing to do—as morally unambiguous and socially responsible—as planting them does today.

It's not at all easy for us to imagine this today, or to find any beauty in a freshly clear-cut landscape, much less a measure of moral satisfaction, but of course many before us have. William James tells a story (retold by Richard Rorty in *Contingency, Irony, and Solidarity*) about traveling in the Appalachian mountains and coming upon a rude homestead that a farmer had recently hacked out of the woods. The man's log cabin, ramshackle garden, and muddy pigpen at first struck James as "hideous, a sort of ulcer," but after the farmer tells him that "we ain't happy here unless we're getting one of those coves under cultivation," he realizes that

> I had been losing the whole inward significance of the situation. Because to me the clearings spoke of naught but denudation, I thought that to those whose sturdy arms and obedient axes had made them they could tell no other story. But when *they* looked on the hideous stumps, what they thought of was personal victory. . . . In short, the clearing which to me was a mere ugly picture on the retina, was to them a symbol redolent with moral memories and sang a very paean of duty, struggle and success.

James's initial impressions of this farmer's homestead echo many eighteenth- and nineteenth-century descriptions of New England farms by English travelers. "The scene is truly savage," one European wrote upon seeing the American landscape for the first time. Early farmers would simply burn or girdle whole forests and then proceed to plant crops amid the leafless trunks and charred stumps. New England fields, most visitors agreed, had "an uncouth and disgusting appearance." Why this repulsion? Probably because by the eighteenth century, Europeans had found themselves with so few trees left that they suddenly acquired a new sense of their value and beauty. The American's attitude toward his virgin forests appalled them as easily as the Brazilian's attitude toward his rain forest appalls us today.

So who is right? Which among the different stories we can tell ourselves about trees and axes is *true*? It is easy and comforting to think that I am more enlightened on the subject of trees than Joe Matyas (or than the Brazilians), but I am beginning to think the truth about trees may be more complicated.

As it happens, the etymology of the word *true* takes us back to the old English word for "tree": a truth, to the Anglo-Saxons, was nothing more than a deeply rooted idea. Just so, my version of a planted tree—envoy to the future, repository of history, index of our respect for the land, spring of aesthetic pleasure, etc.—is "true"; it has deep roots in the culture and seems to serve us well. But as Joe Matyas might have warned me, even the most deeply rooted ideas can fall.

Joe Matyas's tree and mine aren't the only two ever to have shaded this site, of course. I can count perhaps half a dozen other versions of the tree that have found favor in this corner of New England alone, starting with the Indian's. The history of these trees (or tree metaphors) is worth recounting, if only because it suggests how our own truths about the land might someday give way to new and possibly more helpful ones.

Though no Indians are known to have lived in Cornwall, they regularly hunted and traversed the forests here, and most of our roads follow their trails. The Indian landscape was animated by all manner of spirits, and trees were thought to possess venerable souls one was careful not to offend. In the shade of certain trees one found insight. Trees had feelings, eyes, and ears (a notion that has had a certain persistence), and you did not cut one down unless absolutely necessary. Even then, you took the trouble to explain your reasons to the tree and beg its forgiveness.

The American Indians were not the first or the only people ever to consider trees divine; many, if not most, pre-Christian peoples practiced some form of tree worship. Frazer's *Golden Bough* catalogs dozens of instances, from every corner of Northern Europe as well as from Ancient Greece, Rome, and the East. For most of history, in fact, the woods have been thickly populated by spirits and sprites, demons, elves and fairies, and the trees themselves have been regarded as the habitations of gods. (Interestingly, one kind of tree has been revered more widely than any other: the oak, Zeus's tree. The oak's life span may account for this; more even than most trees, oaks transcend men. Frazer suggests another possible reason for the oak's special status: it is the tree most often struck by lightning, and so may be thought to enjoy a special relationship with the heavens.)

If, as has been said, monotheism taught men how they might fear God without fearing his creation, the Puritans took this novel idea to its extreme: they could love God and at the same time detest his creation. The Puritan's tree could scarcely have been more different from the Indian's. In Puritan eyes the New World forest was a "hideous wilderness," "wild and uncouth," a "dismal thicket," in which a person was liable to be lost or killed or, worse still, to fall away from Christ and civilization—to go native. The forest, that shadowy haunt of Satan and uncertainty, deeply offended the Puritans' notions of order and light, indeed of civilization itself. In the captivity narratives that the Puritans (and many later generations of Americans) told themselves to demonize the Indians, the tree is cast as a virtual accomplice to evil: invariably, the red men would tie the captured white woman to one tree and dash her baby's skull against another. To chop down a tree was a supremely righteous act, one by which God's work was advanced and the howling wilderness set back.

Some of the reasons for the Puritans' hard feelings toward trees were surely practical—bringing their form of agriculture to the New World required a herculean labor of deforestation, and righteous hatred of the woods would be a good way to make the work go faster. But it also seems likely that the sight of Indians worshipping trees would have fanned their antipathy; they found themselves in a land where trees were still pagan idols. In chopping them down, the Puritans took their place in an old Christian tradition of animosity toward trees, which the Church justly viewed as rivals; the medieval popes had regularly issued proclamations prohibiting the worship of trees and ordering the destruction of sacred groves. As was often the case when outright prohibition of such a pagan practice failed to eliminate it, Christianity's next move was to co-opt it, and it's possible to interpret the architecture of Gothic cathedrals, whose soaring spaces and filtered light resemble a forest's, as an ingenious attempt to appropriate the sacred grove for Christ.

Though they acted on a more secular authority, the later colonists and then the federalists continued and eventually won the Puritan war against trees.* De-divinized by now, a tree in colonial eyes appeared as

* The best accounts of Puritan and colonial attitudes toward the landscape are William Cronon's ecological history *Changes in the Land* (New York: Hill & Wang, 1983) and John R. Stilgoe's *Common Landscape of America, 1580–1845* (New Haven: Yale University Press, 1982).

either a commodity or a weed. When a colonist looked at a pine tree he saw a ship's mast; in an oak he saw barrel staves. Everything else was in the way. To the colonists deforestation was a synonym for progress. Clearing trees improved the land, and in many cases bolstered one's title to it.

When the land that is now Cornwall was first auctioned off in 500-acre parcels in 1738, the colony stipulated that each new owner had to clear at least six acres of his land within three years or else forfeit title to it. According to town property-tax records, all but a fraction of the town's land had been cleared of trees by 1820. When Joe Matyas bought this hillside in 1919, it would have been virtually bald. But even by then Cornwall's farms had begun to fail and the town's remaining fringe of trees was dilating, spreading down from the untillable hillcrests toward the Housatonic Valley, contesting every clearing in its path. Joe, as one of the last local heirs of the Colonial Tree metaphor, must have struggled mightily to keep the onrushing forest at bay. Like most Americans of the last century, he'd sooner have worshipped an ax than a tree. No doubt he would have approved of Whitman's "Song of the Broad-Axe," in which the ax is depicted as a kind of wellspring from which the new American nation pours forth:

> *The axe leaps!*
> *The solid forest gives fluid utterances,*
> *They tumble forth, they rise and form,*
> *Hut, tent, landing, survey,*
> *Flail, plough, pick, crowbar, spade,*
> *Shingle, rail, prop, wainscot, jamb, lath, panel, gable,*
> *Citadel, ceiling, saloon, academy, organ, exhibition-house,*
> *library. . . .*
> *Capitols of States, and capitol of the nation of States. . . .*
> *The shapes arise!*

The history of tree metaphors is not quite as linear as I am making it out to be. For though the colonial metaphor has held sway on this property for most of its history, elements of other ways of seeing trees have also left their mark, though much more faintly. The two white ash trees, for instance, were probably planted under the influence of a new tree metaphor that arose in England in the seventeenth or eighteenth century, soon after the English awoke to the fact that they had hardly any trees left. The meaning of my ash trees, which I assume were planted by Matyas or his immediate predecessor, is essentially political: standing sentry at the head of the driveway, they declare the bounds of the property and assert the owner's intent to hold onto it indefinitely.

The idea that *planting* trees could have a social or political significance appears to have been invented by the English, though it has since spread widely. According to Keith Thomas's history *Man and the Natural World,* seventeenth- and eighteenth-century aristocrats began planting hardwood trees, usually in lines, to declare the extent of their property and the permanence of their claim to it. "What can be more pleasant," the editor of a magazine for gentlemen asked his readers, "than to have the bounds and limits of your own property preserved and continued from age to age by the testimony of such living and growing witnesses?" Planting trees had the additional advantage of being regarded as a patriotic act, for the Crown had declared a perilous shortage of the hardwood on which the Royal Navy depended.

The Political Tree had been born.

English aristocrats of the time developed an obsession with trees, which they not only planted, but painted and wrote poems about and discussed at tiresome length. (When Washington Irving visited England, he was amused and perplexed to find gentlemen discussing the attributes of individual trees as if they were statues or horses.) Landowners came to identify with their trees, to see in their nobility and rootedness a symbol of their own social standing. Edmund Burke declared that the aristocrats were "the great oaks that shade a country."

This sort of political symbolism is rarely lost on the less-privileged members of the social forest: during the English Revolution, rebels in the countryside made a practice of chopping down the trees on Royalist estates. After the Restoration, replanting trees was regarded as a fitting way for a gentleman to demonstrate his loyalty to the monarchy, and several million hardwoods were planted between 1660 and 1800.

Thus at the same time Americans were hard at work deforesting their continent, the English were embarking on what was probably the first large-scale planting of trees in history. Tree worship had been rediscovered, though its significance now was obviously more social than spiritual. As Thomas points out, tree planting on the scale seen in England in the eighteenth century reflected "not just leisure and a deep purse, but political security and a system of inheritance which gave confidence in the transmission of property. No doubt, this was one of the reasons why it began earlier in England" than elsewhere.

England's great trees, many of which date from this period, not only reflect the country's conservative traditions, they've probably helped to perpetuate them, too. Visiting England early in this century, the Czech writer Karel Čapek speculated that the country's

> splendidly broad-shouldered, ancient, generous, free, vast, venerable trees . . . have had a great influence on Toryism in England. I think they preserve the aristocratic instincts, the historical sense, Conservatism, tariffs, golf, the House of Lords, and other odd and antique things. I should probably be a rabid Radical if I lived in the Street of the Iron Balconies or in the Street of the Grey Bricks, but sitting under an ancient oak tree in the park at Hampton Court I was seriously tempted to acknowledge the value of old things, the high mission of old trees, the harmonious comprehensiveness of tradition, and the legitimacy of esteem for everything that is strong enough to preserve itself for ages.

When men worship trees, Keith Thomas concludes, they are really worshipping their society. Well, *planted* trees anyway—the worship of trees in the wild has a different sense, as we'll see. But the trees planted in England in the eighteenth century are part of that nation's heritage, which might explain why the English were so badly shaken by the October 1987 hurricane.

In our time, the Political Tree has thrived particularly well in the soil of the Middle East. Perhaps because of the movement's English roots, Zionism early on introduced the Political Tree to Palestine, where it remains a highly charged symbol. The Israelis have planted millions of trees in the desert as a way to affirm their claim to the land; they've also made uprooting a tree a crime and required West Bank residents to obtain a permit before planting one on public lands. Naturally the political symbolism of trees has not been lost on the Palestinians, who stunned Israelis in the early days of the *intifada* when they set fire to several Israeli-planted forests. The war of stones has also been a war of trees: the Israeli Defense Forces retaliated by bulldozing Palestinian olive orchards.

In England and America, the Political Tree gradually gave way in the nineteenth century to the Romantic Tree, whose significance is more spiritual than social. This is the tree of Wordsworth, Emerson, Thoreau, and Muir—and of our own time, by and large. "In the woods, we return to reason and faith," Emerson wrote, pointing Americans toward a tree from which they could draw spiritual sustenance—a novel idea in America. In the contemplation and company of the Romantic Tree—self-reliant, abiding, reaching ever heavenward—we could find an antidote to our mean commercial culture, and open ourselves to the infinite. For the tree stands aloof from history, providing a lofty vantage from which we might look beyond its messiness and contingency to "higher laws." Trees "ministered" to men, Thoreau said; they provided for our spiritual and emotional well-being. "Show me two villages, one embowered in trees . . . the other a merely trivial and treeless waste, or with only a single tree or two for suicides, and I shall be sure that in the latter will be found the most starved and bigoted religionists and the most desperate drinkers."

In America, the Romantic Tree and the Colonial Tree coexisted tensely during the second half of the nineteenth century. At about the same time Whitman was celebrating the broadax, Thoreau was composing a tender obituary for a pine tree felled by lumbermen: "A plant which it has taken two centuries to perfect, rising by slow stages into the heavens, had this afternoon ceased to exist. . . . Why does not the village bell sound a knell?" At the turn of the century Gifford Pinchot, Theodore Roosevelt's forester, proposed a new metaphor, the Utilitarian Tree, as a way to reconcile the tree-as-commodity with the tree-as-spiritual-object; under his scheme, trees could still be cut as needed, but judiciously, with an eye toward conserving certain revered stands. Pinchot's compromise didn't hold, however, and eventually Thoreau's tree prevailed over Whitman's ax, at least in the popular mind. Today most of us see the tree, and the forest, through Thoreau's eyes. He would easily have recognized the tree depicted by our nature writers, a tree that stands outside of culture, bearing a kind of moral and spiritual witness. The Romantic Tree is in fact the mirror image of the Puritan Tree: if Puritans clear trees to redeem nature, romantics worship trees to redeem culture. Both regard nature and culture as opponents; they just vote differently.

Obviously the metaphor a people holds about trees will matter greatly to the trees of that time. Puritan Trees tend to get chopped down sanctimoniously. Colonial Trees get chopped down unceremoniously. In stable times Political Trees get planted, but in revolutionary times, they get chopped down—albeit ceremoniously. And the Romantic Tree? Its proper fate is to find itself in a park or wilderness area, out of the way of humans. In general, the Romantic Tree is one you preserve, rather than plant, since much of its spiritual authority derives from its independence from man, its pristine Otherness. And in fact it is to the romantic idea about trees, and nature in general, that we owe the invention of the wilderness area, one of America's great contributions to world culture.

Where do I come down among these metaphors? Somewhere between the Political and the Romantic Tree, I guess. In undertaking to plant, I'm acting in line with the political metaphor: I want to leave my mark here, address the future. But frankly I would have been just as happy to inherit trees, to skip right to romance. I pined for great trees here so that I could entertain Emersonian thoughts in their shade. Most of my unexamined feelings about trees I inherited from the romantics. My initial conclusions about Joe Matyas were exactly the ones Thoreau would have drawn: from this "trivial and treeless waste" he'd have concluded that its owner was a bigoted religionist and desperate drinker—a heathen in a land of nature-worshippers.

But, as I say, I no longer feel comfortable with such a smug characterization. Joe lived by the lights of a different metaphor, one that was necessary to get a certain historical job done: to settle and build America. It was as important a tool in its time as the ax. Though it may be harder for us to see, our own metaphors about nature are no more right or eternal than Joe's. From the standpoint of the next tree metaphor, our own will appear as contingent as Joe's—and probably just as benighted. If the history I've been recounting has anything to teach us, it's that the nineteenth-century idea of a tree standing serenely outside of culture—indeed, the whole idea of nature being "out there," a kind of abiding metaphysical absolute against which we can judge messy, contingent culture—is *itself* a cultural construct, an invention of Emerson and Thoreau and the English romantic poets. A great invention, to be sure—it gave us the wilderness park, our unsurpassed literature of nature, and a lot of terrific camping trips—but we shouldn't mistake it for an eternal verity. Like the Colonial Tree, or the Political Tree, the Romantic Tree is nothing more (or less) than a tool that's proven useful in accomplishing certain important historical tasks.

But I'm beginning to wonder how useful it remains. If I've learned anything in the garden up to now, it is that the romantic's blunt opposition of nature to culture is not helpful. The romantic metaphor offers us no role in nature except as an observer or worshipper; to *act* in nature is to stain it with culture. (Consider the popular usage: land is "virgin" until men "rape" it.) The romantic idea might encourage me to revere and preserve what trees I had, yet it didn't offer much incentive to plant new ones. In fact it is precisely the image of the noble Romantic Tree that made those skinny, price-tagged saplings at the nursery look so pathetic. The political metaphor might be a bit more helpful—it helps keep my eye on a distant prize—but isn't there something just a little presumptuous about those grandiose ideas of planting trees for future generations?

It seems to me we could use a few new tree metaphors about now.

But let me return for a moment from this forest of rather speculative trees to my real one, patiently waiting to be planted.

After the hole was dug, I prepared the soil, a subject on which tree planting authorities are currently divided. The latest thinking holds that, *contra* Snodsmith, the soil should not be improved unless it is unusually poor to begin with—the theory being that a tree growing in a pocket of privileged soil will be spoiled and so fail to develop a sturdy constitution. My own soil is so heavy that I decided to heed the advice of the old school, to lighten and enrich it. So, after loosening the earth in the bottom of the crater with a pitchfork, I added a six-cubic-foot bale of peat moss, a couple of forty-pound bags of composted cow manure, and a few shovelfuls of my own compost (no fertilizer, though: it can burn a young tree's roots). Standing at the bottom of the hole, I turned and stirred the mixture with my pitchfork, and then put some aside for later. I also scored and pitted the walls of the crater; if these are too hard or sheer, the tree is liable to become rootbound, almost as if it were growing in a pot.

The next step is to run a hose and fill the hole with water. This is done not only to ensure an adequate supply of moisture, but to settle the earth and remove any large pockets of air, which could rot any roots exposed to it. I gave the water plenty of time to seep down into the earth and then measured the depth of the hole, using a board and a plumb line. Planting depth is critical: plant the tree too deep and its roots may suffocate; too shallow and they're apt to be exposed. The final grade should barely cover the top of the root ball. If the soil at the bottom of the hole has been disturbed, it's important to make an allowance for settling.

Here at last was a true fifty-dollar hole, a Snodsmithian hole, and the time had come to introduce my maple to it. With the help of a thick wood plank and a few spare hands, I managed to gently work the maple down into its hole without banging up the root ball too badly. Though it's not necessary to remove the burlap (it will decompose soon enough), I did untie the knots around the base of the tree and remove as much of the wire binding the root ball as I could. Then, while Judith held the trunk perpendicular, I began backfilling around the root ball with the prepared soil mixture. After every few spadefuls, I watered and stamped on the fresh ground to squeeze out any air pockets and consolidate the earth's hold. Once the backfill reached the level of the ground, I formed a six-inch lip of soil around the perimeter of the hole to collect rainwater and conduct it toward the roots. I filled this pool with water several times, soaking the roots deeply, and then added a layer of mulch to keep the soil from drying out.

Putting a plant in the ground always seems to diminish it. Suddenly my maple had lost three feet of its height, making it look even more inconsequential than it did standing above ground. And now it was about to get shrimpier still, because John had instructed me to "top it": in order to restore a proper balance between the roots (which have been cut back at the nursery) and the top of the tree, its crown should be thinned out after planting—by as much as a third, some say, though others will argue the point. But if the tree's crown is too large, its diminished root system may not be able to supply enough water to support its profligate leaves, and the tree will go into shock. (This is the reason for planting trees in late fall, when they have no leaves and therefore little need of water; by the time the tree leafs out again in spring, its root system will have recuperated.) So, seeking to right the root/shoot ratio, I climbed a ladder and reluctantly amputated several of my tree's already meager limbs, an act of horticultural mercy I found hard to perform.

The final step is to provide the tree with a measure of protection from the elements during its first year, though even on this point you will get an argument; the latest thinking is against undue coddling of the new tree. I opted for the old bleeding-heart approach anyway, on the theory that conditions on this hillside are especially harsh. To protect its bark from the winter sun and wind, I wrapped the trunk in paper; to thwart the field mice that like to nibble their way around the base of a young tree (thereby killing it), I swaddled its foot in a sock fashioned from a patch of metal window screen. And finally I staked the tree, to keep the wind from disturbing its fragile roots as they make their initial forays into unfamiliar soil.

By the time I stood back to admire my work, dusk had fallen. It was one of those cloudless October evenings when the temperature drops fast, in lockstep with the sun; the night promised a killing frost. And my freshly planted maple, this pricey, twig-topped pole wearing socks and steadied by guy wires, seemed much too vulnerable to spend the night outdoors alone. For all that work, it really didn't look like much—a knobby old guy with a walker out alone on a treeless plain, about as far from our picture of a Romantic Tree as one can imagine. And in the days to follow, I would be disappointed time and again by the failure of visitors even to notice my tree without prompting, let alone admire it. But the longer I stood there gazing at my tree, the more of it I could see. Maybe it was the late, uncertain light, but after a while I had little trouble imagining its future taking shape. I looked at the thin, knuckled twigs and could picture them, as if in time-lapse, leafing out and branching, spring after spring, one branch into two into four into eight into sixteen, my tree compounding itself every summer in a geometrical progression that blossomed at last into a massive oval crown.

From my desk up in the barn loft I have a good view of the new tree, and whenever my attention wanders from my work, it seems to settle there, amid its leafless branches. A frail thing to burden with so much reflection, I know, but that seems to be the fate of trees in a world of humans—our thoughts and metaphors cling to them like iron filings to a magnet. Obviously trees exist apart from our evolving pictures of them—we didn't *invent* them—but trees were married to our metaphors so long ago, we have no idea what they would be like single. Every time we think we've figured out what a tree *really* is—the habitation of gods, a commodity, part and parcel of transcendent nature, component of the forest ecosystem—it turns out we've simply come up with a temporarily handy new description of it. Yet given who we are, that's no mean thing: our metaphors matter. Indeed, our metaphors about trees by and large determine the fate of trees.

Trees have been in the news quite a bit lately. Scientists warn that they are in trouble, and that their health is bound up with our own in ways never imagined before. Deforestation may be contributing to potentially catastrophic changes in the Earth's atmosphere. It's no wonder, then, that images of trees seem to be showing up everywhere just now: in art galleries, on magazine covers, in product logos and advertisements, in the speeches of politicians. My hunch is that we sense our old metaphors about trees, and nature as a whole, are wearing thin, and we're casting around for new and more powerful ones. By the time my maple reaches its maturity, it probably will mean something very different from what it means today.

What might these new metaphors be? Some philosophers and activists have recently advanced the notion that my tree (and nature generally) possesses "rights." They see Western history as a continuing struggle to widen the circle of rights-holders, from nobles to property holders to white males to men generally and, most recently, to women. They propose we now draw this circle still wider, to encompass nature. With perfectly straight faces, they offer analogies between the condition of African-Americans prior to abolition and the condition of nature today. From this equation flows a justification for radical action in defense of nature, and groups like Earth First! have made a practice of spiking trees to defend them from loggers. Once you accept that trees have equal rights, the fact that spiking them endangers the lives of loggers and mill operators will seem a lot less troubling.

One legal scholar, Christopher D. Stone, has gone so far as to argue in a book entitled *Should Trees Have Standing?* that forests, lakes, and mountains should be granted the right to sue (called "standing") in American courts. The idea is not quite as farfetched as it sounds; corporations and ships are already "persons" in the eyes of the law, so why not also trees? Stone's argument was actually accepted by Justice William O. Douglas, and in recent years a handful of suits to protect natural areas have been successfully filed on behalf of trees and other natural objects.

I'm not sure I like the ideas of my tree growing up to be litigious. Though the proponents of nature's rights surely have the best interests of my tree and the rest of nature at heart, I worry that a world in which trees have rights would probably be a world in which human rights have been substantially diluted. The rights of the individual, such a hard-won and tenuous achievement of Western history, would not fare well in a world of "natural rights," if only because, in nature, species always count for more than individuals. From the "biocentric" perspective that radical environmentalists are pressing us to adopt, the last few grizzlies matter more than any given human. In seeking to expand liberalism to encompass nature, we could end up wrecking liberalism.

This is merely a pragmatic objection, of course, and it won't cut much ice with people who think they've discovered a new truth about nature. But of course the idea of the tree as rights holder is really nothing more than another metaphor, one we are free to accept or reject. If it does catch on in this country (and I worry it might), that will be because it happens to align with our liberal tradition, as well as with Thoreau's Romantic Tree. (For what is the Litigious Tree but a Romantic Tree with a lawyer?) Yet for all their talk of biocentrism, proponents of nature's rights never really escape the trap of anthropocentrism: rights, after all, are a human invention, ours to grant or withhold.

Still, can't we come up with a metaphor less awkward than yet another one based on "rights"? In fact, science has recently proposed some new descriptions of trees that strike me as much more promising, and which, in retrospect, lend mankind's old, strong feelings about trees an eerie prescience.

Think of the tree as the Earth's breathing apparatus, an organ that helps regulate the planet's atmosphere by exhaling fresh oxygen and inhaling the carbon that animals, decay, and civilization spew into it. The tree, under this new description, is not merely a member of the local forest ecosystem (where we've known for some time that it exerts considerable influence on the local life, soil, and even climate); it's also a vital organ in a global system more intricate and interdependent than we ever realized. The Earth may be not a spaceship but an organism, and the trees may be its lungs.

Using instruments of gas analysis set up on the slope of a volcano in Hawaii, mankind has now actually observed the breathing of the earth, which follows an annual rhythm: every summer, the quantity of carbon dioxide in the atmosphere of the northern hemisphere falls as the forests inhale; every winter, after photosynthesis subsides and civilization steps up its combustion of fossil fuel, the carbon dioxide levels rise again, a little higher each year. (In our time, the Earth's breathing may be growing labored, as the forest's inhalation of carbon dioxide struggles to keep up with the hot, heavy breath of civilization.) Here, then, are the lineaments of a new tree metaphor, one of great force, beauty, and import.

Science has also come to regard trees as barometers of our ecological health, since they seem to exhibit the effects of the damage man is doing to the environment long before they show up elsewhere. Ecologists think that the greenhouse effect will show up first in the forests, where cool-weather tree species, unable to migrate northward fast enough to keep pace with a warming climate, may soon begin to sicken and die. Already forests in New England show the effects of acid rain. (The reason, you will recall, that I ended up with a Norway maple in the first place; one thing my tree will probably signify is our early efforts to adapt to this new world.) Trees are like the canaries miners used to bring with them into the coal mines; since the birds succumbed to poisonous gases long before humans did, they warned miners of unseen dangers.

Given the choice, I would much prefer to see the Lung Tree or the Canary Tree catch on than the Litigious Tree. These first two metaphors (which are in fact closely related) have the virtue of forcing us to see the connections between our small, local acts and the health of the planet. They encourage us to preserve what trees we have and to plant new ones. But even more important, I think, is that the lung metaphor puts us in a reciprocal relation with the trees once again. It undercuts romantic notions of their Otherness, pointing us toward an existential plane we share. If we come to think of trees as lungs and the Earth as an organism, it will no longer make sense to think of ourselves as being outside nature, or even to think of trees as being outside culture. Indeed, the whole inside/outside metaphor might wither away, and that would be a good thing.

It's obviously impossible to predict which, if any, of these new metaphors will catch on. That will depend on how useful they prove, as well as on the usual vicissitudes of our ongoing conversation about nature. For a new Thoreau, who this time might or might not be a scientist, could come along at any time and remake the tree entirely, along lines we can't possibly foresee. I do know this, though: if I could have news of my maple one hundred years from now, I would know a great deal about nature's fate.

One day not long ago, I gave some thought to exactly what sort of news of my tree I would want. It was early in the morning after a night that had brought the season's first snow. The sun was so low in the eastern sky, and so bright, that the maple cast an uncommonly long and sharp-edged shadow on the fresh page of snow. It raced straight west across the meadow, angled up a small hill, and then shot off deep into the woods, where I lost track of it.

So what did I want from there, up ahead? Certainly a botanist's report on my tree's health would be useful. The Norway maple is a cool-weather species, and if it has sickened in the heat of 2091, I will know that the greenhouse effect was real and that we did not avert it. But perhaps even more revealing than a scientist's account would be to have a letter from that time, one that happened to devote a few sentences of description to my tree, in everyday language. From that I might learn how people in 2091 looked at a tree, and this would pretty much tell me how nature was faring then. If the letter described the tree in terms that would have been familiar to Joe Matyas—or Henry Thoreau, for that matter—that would be cause for worry, for it would mean we'd gotten mired in old metaphors about nature, and had probably failed to extricate ourselves from our predicament.

But maybe the letter would bring evidence of a new metaphor, something vivid and powerful and, for a time at least, *true.* At first it would probably seem strange, even incomprehensible. But eventually its sense would dawn. *So that's what a tree is! How could we ever have thought otherwise?* There might then be reason to hope that some new truth had put down roots, that perhaps we had put our relationship to nature on a sounder footing at last.

CHAPTER 10

The Idea of a Garden

The biggest news to come out of my town in many years was the tornado, or tornadoes, that careened through here on July 10, 1989, a Monday. Shooting down the Housatonic River Valley from the Berkshires, it veered east over Coltsfoot Mountain and then, after smudging the sky a weird gray green, proceeded to pinball madly from hillside to hillside for about fifteen minutes before wheeling back up into the sky. This was part of the same storm that ripped open the bark of my ash tree. But the damage was much, much worse on the other side of town. Like a gigantic, skidding pencil eraser, the twister neatly erased whole patches of woods and roughly smeared many other ones, where it wiped out just the tops of the trees. Overnight, large parts of town were rendered unrecognizable.

One place where the eraser came down squarely was in the Cathedral Pines, a famous forest of old-growth white pine trees close to the center of town. A kind of local shrine, this forty-two-acre forest was one of the oldest stands of white pine in New England, the trees untouched since about 1800. To see it was to have some idea how the New World forest must have looked to the first settlers, and in 1985 the federal government designated it a "national natural landmark." To enter Cathedral Pines on a hot summer day was like stepping out of the sun into a dim cathedral, the sunlight cooled and sweetened by the trillions of pine needles as it worked its way down to soft, sprung ground that had been unacquainted with blue sky for the better part of two centuries. The storm came through at about five in the evening, and it took only a few minutes of wind before pines more than one hundred fifty feet tall and as wide around as missiles lay jackstrawed on the ground like a fistful of pencils dropped from a great height. The wind was so thunderous that people in houses at the forest's edge did not know trees had fallen until they ventured outside after the storm had passed. The following morning, the sky now clear, was the first in more than a century to bring sunlight crashing down onto this particular patch of earth.

"It is a terrible mess," the first selectman told the newspapers; "a tragedy," said another Cornwall resident, voicing the deep sense of loss shared by many in town. But in the days that followed, the selectman and the rest of us learned that our responses, though understandable, were shortsighted, unscientific, and, worst of all, anthropocentric. "It may be a calamity to us," a state environmental official told a reporter from the *Hartford Courant*, but "to biology it is not a travesty. It is just a natural occurrence." The Nature Conservancy, which owns Cathedral Pines, issued a press release explaining that "Monday's storm was just another link in the continuous chain of events that is responsible for shaping and changing this forest."

It wasn't long before the rub of these two perspectives set off a controversy heated enough to find its way into the pages of *The New York Times.* The Nature Conservancy, in keeping with its mandate to maintain its lands in a "state of nature," indicated that it would leave Cathedral Pines alone, allowing the forest to take its "natural course," whatever that might be. To town officials and neighbors of the forest this was completely unacceptable. The downed trees, besides constituting an eyesore right at the edge of town, also posed a fire hazard. A few summers of drought, and the timber might go up in a blaze that would threaten several nearby homes and possibly even the town itself. Many people in Cornwall wanted Cathedral Pines cleared and replanted, so that at least the next generation might live to see some semblance of the old forest.

A few others had the poor taste to point out the waste of more than a million board-feet of valuable timber, stupendous lengths of unblemished, knot-free pine.

The newspapers depicted it as a classic environmental battle, pitting the interests of man against nature, and in a way it was that. On one side were the environmental purists, who felt that *any* intervention by man in the disposition of this forest would be unnatural. "If you're going to clean it up," one purist declared in the local press, "you might as well put up condos." On the other side stood the putative interests of man, variously expressed in the vocabulary of safety (the fire hazard), economics (the wasted lumber), and aesthetics (the "terrible mess").

Everybody enjoys a good local fight, but I have to say I soon found the whole thing depressing. This was indeed a classic environmental battle, in that it seemed to exemplify just about everything that's wrong with the way we approach problems of this kind these days. Both sides began to caricature each other's positions: the selectman's "terrible mess" line earned him ridicule for his anthropocentrism in the letters page of *The New York Times;* he in turn charged a Yale scientist who argued for noninterference with "living in an ivory tower."

But as far apart as the two sides seemed to stand, they actually shared more common ground than they realized. Both started from the premise that man and nature were irreconcilably opposed, and that the victory of one necessarily entailed the loss of the other. Both sides, in other words, accepted the premises of what we might call the "wilderness ethic," which is based on the assumption that the relationship of man and nature resembles a zero-sum game. This idea, widely held and yet largely unexamined, has set the terms of most environmental battles in this country since the very first important one: the fight over the building of the Hetch Hetchy Dam in 1907, which pitted John Muir against Gifford Pinchot, whom Muir used to call a "temple destroyer." Watching my little local debate unfold over the course of the summer, and grow progressively more shrill and sterile, I began to wonder if perhaps the wilderness ethic itself, for all that it has accomplished in this country over the past century, had now become part of the problem. I also began to wonder if it might be possible to formulate a different ethic to guide us in our dealings with nature, at least in some places some of the time, an ethic that would be based not on the idea of wilderness but on the idea of a garden.*

Foresters who have examined sections of fallen trees in Cathedral Pines think that the oldest trees in the forest date from 1780 or so, which suggests that the site was probably logged by the first generation of settlers. The Cathedral Pines are not, then, "virgin growth." The rings of felled trees also reveal a significant growth spurt in 1840, which probably indicates that loggers removed hardwood trees in that year, leaving the pines to grow without competition. In 1883, the Calhouns, an old Cornwall family whose property borders the forest, bought the land to protect the trees from the threat of logging; in 1967 they deeded it to the Nature Conservancy, stipulating that it be maintained in its natural state. Since then, and up until the tornado made its paths impassable, the forest has been a popular place for hiking and Sunday outings. Over the years, more than a few Cornwall residents have come to the forest to be married.

Cathedral Pines is not in any meaningful sense a wilderness. The natural history of the forest intersects at many points with the social history of Cornwall. It is the product of early logging practices, which clear-cut the land once and then cut it again, this time selectively, a hundred years later. Other human factors almost certainly played a part in the forest's history; we can safely assume that any fires in the area were extinguished before they reached Cathedral Pines. (Though we don't

---

* In developing some of the ideas for this chapter, I've drawn from a panel discussion on environmental ethics that I moderated for the April 1990 issue of *Harper's Magazine.* The participants were James Lovelock, Frederick Turner, Daniel Botkin, Dave Foreman, and Robert Yaro. This chapter also owes a lot to the work of Wendell Berry, René Dubos, William Cronon, William Jordan III, and Alston Chase.

ordinarily think of it in these terms, fire suppression is one of the more significant effects that the European has had on the American landscape.) Cathedral Pines, then, is in some part a man-made landscape, and it could reasonably be argued that to exclude man at this point in its history would constitute a break with its past.

But both parties to the dispute chose to disregard the actual history of Cathedral Pines, and instead to think of the forest as a wilderness in the commonly accepted sense of that term: a pristine place untouched by white men. Since the romantics, we've prized such places as refuges from the messiness of the human estate, vantages from which we might transcend the vagaries of that world and fix on what Thoreau called "higher laws." Certainly an afternoon in Cathedral Pines fostered such feelings, and its very name reflects the pantheism that lies behind them. Long before science coined the term *ecosystem* to describe it, we've had the sense that nature undisturbed displays a miraculous order and balance, something the human world can only dream about. When man leaves it alone, nature will tend toward a healthy and abiding state of equilibrium. Wilderness, the purest expression of this natural law, stands out beyond history.

These are powerful and in many ways wonderful ideas. The notion of wilderness is a kind of taboo in our culture, in many cases acting as a check on our inclination to dominate and spoil nature. It has inspired us to set aside such spectacular places as Yellowstone and Yosemite. But wilderness is also a profoundly alienating idea, for it drives a large wedge between man and nature. Set against the foil of nature's timeless cycles, human history appears linear and unpredictable, buffeted by time and chance as it drives blindly into the future. Natural history, by comparison, obeys fixed and legible laws, ones that make the "laws" of human history seem puny, second-rate things scarcely deserving of the label. We have little idea what the future holds for the town of Cornwall, but surely nature has a plan for Cathedral Pines; leave the forest alone and that plan—which science knows by the name of "forest succession"—will unfold inexorably, in strict accordance with natural law. A new climax forest will emerge as nature works to restore her equilibrium—or at least that's the idea.

The notion that nature has a plan for Cathedral Pines is a comforting one, and certainly it supplies a powerful argument for leaving the forest alone. Naturally I was curious to know what that plan was: what does nature do with an old pine forest blown down by a tornado? I consulted a few field guides and standard works of forest ecology hoping to find out.

According to the classical theory of forest succession, set out in the nineteenth century by, among others, Henry Thoreau, a pine forest that has been abruptly destroyed will usually be succeeded by hardwoods, typically oak. This is because squirrels commonly bury acorns in pine forests and neglect to retrieve many of them. The oaks sprout and, because shade doesn't greatly hinder young oaks, the seedlings frequently manage to survive beneath the dark canopy of a mature pine forest. Pine seedlings, on the other hand, require more sunlight than a mature pine forest admits; they won't sprout in shade. So by the time the pine forest comes down, the oak saplings will have had a head start in the race to dominate the new forest. Before any new pines have had a chance to sprout, the oaks will be well on their way to cornering the sunlight and inheriting the forest.

This is what I read, anyway, and I decided to ask around to confirm that Cathedral Pines was expected to behave as predicted. I spoke to a forest ecologist and an expert on the staff of the Nature Conservancy. They told me that the classical theory of pine-forest succession probably does describe the underlying tendency at work in Cathedral Pines. But it turns out that a lot can go, if not "wrong" exactly, then at least differently. For what if there are no oaks nearby? Squirrels will travel only so far in search of a hiding place for their acorns. Instead of oaks, there may be hickory nuts stashed all over Cathedral Pines. And then there's the composition of species planted by the forest's human neighbors to consider; one of these, possibly some exotic (that is, non-native), could conceivably race in and take over.

"It all depends," is the refrain I kept hearing as I tried to pin down nature's intentions for Cathedral Pines. Forest succession, it seems, is only a theory, a metaphor of our making, and almost as often as not nature makes a fool of it. The number of factors that will go into the determination of Cathedral Pines' future is almost beyond comprehension. Consider just this small sample of the things that could happen to alter irrevocably its future course:

A lightning storm—or a cigarette butt flicked from a passing car—ignites a fire next summer. Say it's a severe fire, hot enough to damage the fertility of the soil, thereby delaying recovery of the forest for decades. Or say it rains that night, making the fire a mild one, just hot enough to kill the oak saplings and allow the relatively fire-resistant pine seedlings to flourish without competition. A new pine forest after all? Perhaps. But what if the population of deer happens to soar the following year? Their browsing would wipe out the young pines and create an opening for spruce, the taste of which deer happen not to like.

Or say there is no fire. Without one, it could take hundreds of years for the downed pine trees to rot and return their nutrients to the soil. Trees grow poorly in the exhausted soil, but the seeds of brambles, which can lie dormant in the ground for fifty years, sprout and proliferate: we end up with a hundred years of brush. Or perhaps a breeze in, say, the summer of 1997 carries in seedpods from the Norway maple standing in a nearby front yard at the precise moment when conditions for their germination are perfect. Norway maple, you'll recall, is a European species, introduced here early in the nineteenth century and widely planted as a street tree. Should this exotic species happen to prevail, Cathedral Pines becomes one very odd-looking and awkwardly named wilderness area.

But the outcome could be much worse. Let's say the rains next spring are unusually heavy, washing all the topsoil away (the forest stood on a steep hillside). Only exotic weed species can survive now, and one of these happens to be Japanese honeysuckle, a nineteenth-century import of such rampant habit that it can choke out the growth of all trees indefinitely. We end up with no forest at all.

Nobody, in other words, can say what will happen in Cathedral Pines. And the reason is not that forest ecology is a young or imperfect science, but because *nature herself doesn't know what's going to happen here.* Nature has no grand design for this place. An incomprehensibly various and complex set of circumstances—some of human origin, but many not—will determine the future of Cathedral Pines. And whatever that future turns out to be, it would not unfold in precisely the same way twice. Nature may possess certain inherent tendencies, ones that theories such as forest succession can describe, but chance events can divert her course into an almost infinite number of different channels.

It's hard to square this fact with our strong sense that some kind of quasi-divine order inheres in nature's workings. But science lately has been finding that contingency plays nearly as big a role in natural history as it does in human history. Forest ecologists today will acknowledge that succession theories are little more than comforting narratives we impose on a surprisingly unpredictable process; even so-called climax forests are sometimes superseded. (In many places in the northern United States today, mature stands of oak are inexplicably being invaded by maples—skunks at the climax garden party.) Many ecologists will now freely admit that even the concept of an ecosystem is only a metaphor, a human construct imposed upon a much more variable and precarious reality. An ecosystem may be a useful concept, but no ecologist has ever succeeded in isolating one in nature. Nor is the process of evolution as logical or inexorable as we have thought. The current thinking in paleontology holds that the evolution of any given species, our own included, is not the necessary product of any natural laws, but rather the outcome of a concatenation of chance events—of "just history" in the words of Stephen Jay Gould. Add or remove any single happenstance—the asteroid fails to wipe out the dinosaurs; a little chordate worm called *Pikaia* succumbs in the Burgess extinction—and humankind never arrives.

Across several disciplines, in fact, scientists are coming to the conclusion that more "just history" is at work in nature than had previously been thought. Yet our metaphors still picture nature as logical, stable, and ahistorical—more like a watch than, say, an organism or a stock exchange, to name two metaphors that may well be more apt. Chance and contingency, it turns out, are everywhere in nature; she has no fixed goals, no unalterable pathways into the future, no inflexible rules that she herself can't bend or break at will. She is more like us (or we are more like her) than we ever imagined.

To learn this, for me at least, changes everything. I take it to be profoundly good news, though I can easily imagine how it might trouble some people. For many of us, nature is a last bastion of certainty; wilderness, as something beyond the reach of history and accident, is one of the last in our fast-dwindling supply of metaphysical absolutes, those comforting transcendental values by which we have traditionally taken our measure and set our sights. To take away predictable, divinely ordered nature is to pull up one of our last remaining anchors. We are liable to float away on the trackless sea of our own subjectivity.

But the discovery that time and chance hold sway even in nature can also be liberating. Because contingency is an invitation to participate in history. Human choice is unnatural only if nature is deterministic; human change is unnatural only if she is changeless in our absence. If the future of Cathedral Pines is up for grabs, if its history will always be the product of myriad chance events, then why shouldn't we also claim our place among all those deciding factors? For aren't we also one of nature's contingencies? And if our cigarette butts and Norway maples and acid rain are going to shape the future of this place, then why not also our hopes and desires?

Nature will condone an almost infinite number of possible futures for Cathedral Pines. Some would be better than others. True, what we would regard as "better" is probably not what the beetles would prefer. But nature herself has no strong preference. That doesn't mean she will countenance *any* outcome; she's already ruled out many possible futures (tropical rain forest, desert, etc.) and, all things being equal, she'd probably lean toward the oak. But all things aren't equal (*her* idea) and she is evidently happy to let the free play of numerous big and little contingencies settle the matter. To exclude from these human desire would be, at least in this place at this time, arbitrary, perverse and, yes, unnatural.

Establishing that we should have a vote in the disposition of Cathedral Pines is much easier than figuring out how we should cast it. The discovery of contingency in nature would seem to fling open a Pandora's box. For if there's nothing fixed or inevitable about nature's course, what's to stop us from concluding that anything goes? It's a whole lot easier to assume that nature left to her own devices knows what's best for a place, to let ourselves be guided by the wilderness ethic.

And maybe that's what we should do. Just because the wilderness ethic is based on a picture of nature that is probably more mythical than real doesn't necessarily mean we have to discard it. In the same way that the Declaration of Independence begins with the useful fiction that "all men are created equal," we could simply stipulate that Cathedral Pines *is* wilderness, and proceed on that assumption. The test of the wilderness ethic is not how truthful it is, but how useful it is in doing what we want to do—in protecting and improving the environment.

So how good a guide is the wilderness ethic in this particular case? Certainly treating Cathedral Pines as a wilderness will keep us from building condos there. When you don't trust yourself to do the right thing, it helps to have an authority as wise and experienced as nature to decide matters for you. But what if nature decides on Japanese honeysuckle—three hundred years of wall-to-wall brush? We would then have a forest not only that we don't like, but that isn't even a wilderness, since it was man who brought Japanese honeysuckle to Cornwall. At this point in history, after humans have left their stamp on virtually every corner of the Earth, doing nothing is frequently a poor recipe for wilderness. In many cases it leads to a gradually deteriorating environment (as seems to be happening in Yellowstone), or to an environment shaped in large part by the acts and mistakes of previous human inhabitants.

If it's real wilderness we want in Cathedral Pines, and not merely an imagined innocence, we will have to restore it. This is the paradox faced by the Nature Conservancy and most other advocates of wilderness: at this point in history, creating a landscape that bears no marks of human intervention will require a certain amount of human intervention. At a minimum it would entail weeding the exotic species from Cathedral Pines, and that is something the Nature Conservancy's strict adherence to the wilderness ethic will not permit.

But what if the Conservancy *was* willing to intervene just enough to erase any evidence of man's presence? It would soon run up against some difficult questions for which its ethic leaves it ill-prepared. For what is the "real" state of nature in Cathedral Pines? Is it the way the forest looked before the settlers arrived? We could restore that condition by removing all traces of European man. Yet isn't that a rather Eurocentric (if not racist) notion of wilderness? We now know that the Indians were not the ecological eunuchs we once thought. They too left their mark on the land: fires set by Indians determined the composition of the New England forests and probably created that "wilderness" we call the Great Plains. For true untouched wilderness we have to go a lot further back than 1640 or 1492. And if we want to restore the landscape to its pre-Indian condition, then we're going to need a lot of heavy ice-making equipment (not to mention a few woolly mammoths) to make it look right.

But even that would be arbitrary. In fact there is no single moment in time that we can point to and say, *this* is the state of nature in Cathedral Pines. Just since the last ice age alone, that "state of nature" has undergone a thorough revolution every thousand years or so, as tree species forced south by the glaciers migrated back north (a process that is still going on), as the Indians arrived and set their fires, as the large mammals disappeared, as the climate fluctuated—as all the usual historical contingencies came on and off the stage. For several thousand years after the ice age, this part of Connecticut was a treeless tundra; is *that* the true state of nature in Cathedral Pines? The inescapable fact is that, if we want wilderness here, we will have to choose *which* wilderness we want—an idea that is inimical to the wilderness ethic. For wasn't the attraction of wilderness precisely the fact that it relieved us of having to make choices—wasn't nature going to decide, letting us off the hook of history and anthropocentrism?

No such luck, it seems. "Wilderness" is not nearly as straightforward or dependable a guide as we'd like to believe. If we do nothing, we may end up with an impoverished weed patch of our own (indirect) creation, which would hardly count as a victory for wilderness. And if we want to restore Cathedral Pines to some earlier condition, we're forced into making the kinds of inevitably anthropocentric choices and distinctions we turned to wilderness to escape. (Indeed, doing a decent job of wilderness restoration would take all the technology and scientific know-how humans can muster.) Either way, there appears to be no escape from history, not even in nature.

The reason that the wilderness ethic isn't very helpful in a place like Cathedral Pines is that it's an absolutist ethic: man or nature, it says, pick one. As soon as history or circumstance blurs that line, it gets us into trouble. There are times and places when man or nature is the right and necessary choice; back at Hetch Hetchy in 1907 that may well have been the case. But it seems to me that these days most of the environmental questions we face are more like the ambiguous ones posed by Cathedral Pines, and about these the wilderness ethic has less and less to say that is of much help.

The wilderness ethic doesn't tell us what to do when Yellowstone's ecosystem begins to deteriorate, as a result not of our interference but of our neglect. When a species threatens to overwhelm and ruin a habitat because history happened to kill off the predator that once kept its population in check, the ethic is mute. It is confounded, too, when the only hope for the survival of another species is the manipulation of its natural habitat by man. It has nothing to say in all those places where development is desirable or unavoidable except: Don't do it. When we're forced to choose between a hydroelectric power plant and a nuclear one, it refuses to help. That's because the wilderness ethic can't make distinctions between one kind of intervention in nature and another—between weeding Cathedral Pines and developing a theme park there. "You might as well put up condos" is its classic answer to any plan for human intervention in nature.

"All or nothing," says the wilderness ethic, and in fact we've ended up with a landscape in America that conforms to that injunction remarkably well. Thanks to exactly this kind of either/or thinking, Americans have done an admirable job of drawing lines around certain sacred areas (we did invent the wilderness area) and a terrible job of managing the rest of our land. The reason is not hard to find: the only environmental ethic we have has nothing useful to say about those areas outside the line. Once a landscape is no longer "virgin" it is typically written off as fallen, lost to nature, irredeemable. We hand it over to the jurisdiction of that other sacrosanct American ethic: laissez-faire economics. "You might as well put up condos." And so we do.

Indeed, the wilderness ethic and laissez-faire economics, antithetical as they might at first appear, are really mirror images of one another. Each proposes a quasi-divine force—Nature, the Market—that, left to its own devices, somehow knows what's best for a place. Nature and the market are both self-regulating, guided by an invisible hand. Worshippers of either share a deep, Puritan distrust of man, taking it on faith that human tinkering with the natural or economic order can only pervert it. Neither will acknowledge that their respective divinities can also err: that nature produces the AIDS virus as well as the rose, that the same markets that produce stupendous wealth can also crash. (Actually, worshippers of the market are a bit more realistic than worshippers of nature: they long ago stopped relying on the free market to supply us with such necessities as food and shelter. Though they don't like to talk about it much, they accept the need for society to "garden" the market.)

Essentially, we have divided our country in two, between the kingdom of wilderness, which rules about 8 percent of America's land, and the kingdom of the market, which rules the rest. Perhaps we should be grateful for secure borders. But what do those of us who care about nature do when we're on the market side, which is most of the time? How do we behave? What are our goals? We can't reasonably expect to change the borders, no matter how many power lines and dams Earth First! blows up. No, the wilderness ethic won't be of much help over here. Its politics are bound to be hopelessly romantic (consisting of impractical schemes to redraw the borders) or nihilistic. Faced with hard questions about how to confront global environmental problems such as the greenhouse effect or ozone depletion (problems that respect no borders), adherents of the wilderness ethic are apt to throw up their hands in despair and declare the "end of nature."

The only thing that's really in danger of ending is a romantic, pantheistic idea of nature that we invented in the first place, one whose passing might well turn out to be a blessing in disguise. Useful as it has been in helping us protect the sacred 8 percent, it nevertheless has failed to prevent us from doing a great deal of damage to the remaining 92 percent. This old idea may have taught us how to worship nature, but it didn't tell us how to live with her. It told us more than we needed to know about virginity and rape, and almost nothing about marriage. The metaphor of divine nature can admit only two roles for man: as worshipper (the naturalist's role) or temple destroyer (the developer's). But that drama is all played out now. The temple's been destroyed—if it ever was a temple. Nature *is* dead, if by nature we mean something that stands apart from man and messy history. And now that it is, perhaps we can begin to write some new parts for ourselves, ones that will show us how to start out from here, not from some imagined state of innocence, and let us get down to the work at hand.

Thoreau and Muir and their descendants went to the wilderness and returned with the makings of America's first environmental ethic. Today it still stands, though somewhat strained and tattered. What if now, instead of to the wilderness, we were to look to the garden for the makings of a new ethic? One that would not necessarily supplant the earlier one, but might give us something useful to say in those cases when it is silent or unhelpful?

It will take better thinkers than me to flesh out what such an ethic might look like. But even my limited experience in the garden has persuaded me that the materials needed to construct it—the fresh metaphors about nature we need—may be found there. For the garden is a place with long experience of questions having to do with man *in* nature. Below are some provisional notes, based on my own experiences and the experiences of other gardeners I've met or read, on the kinds of answers the garden is apt to give.

1. An ethic based on the garden would give local answers. Unlike the wilderness idea, it would propose different solutions in different places and times. This strikes me as both a strength and a weakness. It's a weakness because a garden ethic will never speak as clearly or univocally as the wilderness ethic does. In a country as large and geographically various as this, it is probably inevitable that we will favor abstract landscape ideas—grids, lawns, monocultures, wildernesses—which can be applied across the board, even legislated nationally; such ideas have the power to simplify and unite. Yet isn't this power itself part of the problem? The health of a place generally suffers whenever we impose practices on it that are better suited to another place; a lawn in Virginia makes sense in a way that a lawn in Arizona does not.

So a garden ethic would begin with Alexander Pope's famous advice to landscape designers: "Consult the Genius of the Place in all." It's hard to imagine this slogan ever replacing Earth First!'s "No Compromise in Defense of Mother Earth" on American bumper stickers; nor should it, at least not everywhere. For Pope's dictum suggests that there are places whose "genius" will, if hearkened to, counsel "no compromise." Yet what is right for Yosemite is not necessarily right for Cathedral Pines.

2. The gardener starts out from here. By that I mean, he accepts contingency, his own and nature's. He doesn't spend a lot of time worrying about whether he has a god-given right to change nature. It's enough for him to know that, for some historical or biological reason, humankind finds itself living in places (six of the seven continents) where it must substantially alter the environment in order to survive. If we had remained on African savanna things might be different. And if I lived in zone six I could probably grow good tomatoes without the use of plastic. The gardener learns to play the hand he's been dealt.

3. A garden ethic would be frankly anthropocentric. As I began to understand when I planted my roses and my maple tree, we know nature only through the screen of our metaphors; to see her plain is probably impossible. (And not necessarily desirable, as George Eliot once suggested: "If we could hear the squirrel's heartbeat, the sound of the grass growing, we should die of that roar." Without the editing of our perceptions, nature might prove unbearable.) Melville was describing all of nature when he described the whiteness of the whale, its "dumb blankness, full of meaning." Even wilderness, in both its satanic and benevolent incarnations, is an historical, man-made idea. Every one of our various metaphors for nature—"wilderness," "ecosystem," "Gaia," "resource," "wasteland"—is already a kind of garden, an indissoluble mixture of our culture and whatever it is that's really out there. "Garden" may sound like a hopelessly anthropocentric concept, but it's probably one we can't get past.

The gardener doesn't waste much time on metaphysics—on figuring out what a "truer" perspective on nature (such as biocentrism or geocentrism) might look like. That's probably because he's noticed that most of the very long or wide perspectives we've recently been asked to adopt (including the one advanced by the Nature Conservancy in Cathedral Pines) are indifferent to our well-being and survival as a species. On this point he agrees with Wendell Berry—that "it is not natural to be disloyal to one's own kind."

4. That said, though, the gardener's conception of his self-interest is broad and enlightened. Anthropocentric as he may be, he recognizes that he is dependent for his health and survival on many other forms of life, so he is careful to take their interests into account in whatever he does. He is in fact a wilderness advocate of a certain kind. It is when he respects and nurtures the wilderness of his soil and his plants that his garden seems to flourish most. Wildness, he has found, resides not only out there, but right here: in his soil, in his plants, even in himself. Overcultivation tends to repress this quality, which experience tells him is necessary to health in all three realms. But wildness is more a quality than a place, and though humans can't manufacture it, they can nourish and husband it. That is precisely what I'm doing when I make compost and return it to the soil; it is what we could be doing in Cathedral Pines (and not necessarily by leaving the place alone). The gardener cultivates wildness, but he does so carefully and respectfully, in full recognition of its mystery.

5. The gardener tends not to be romantic about nature. What could be more natural than the storms and droughts and plagues that ruin his garden? Cruelty, aggression, suffering—these too are nature's offspring (and not, as Rousseau tried to convince us, culture's). Nature is probably a poor place to look for values. She was indifferent to humankind's arrival, and she is indifferent to our survival.

It's only in the last century or so that we seem to have forgotten this. Our romance of nature is a comparatively recent idea, the product of the industrial age's novel conceit that nature could be conquered, and probably also of the fact that few of us work with nature directly anymore. But should current weather forecasts prove to be accurate (a rapid, permanent warming trend accompanied by severe storms), our current romance will look like a brief historical anomaly, a momentary lapse of judgment. Nature may once again turn dangerous and capricious and unconquerable. When this happens, we will quickly lose our crush on her.

Compared to the naturalist, the gardener never fell head over heels for nature. He's seen her ruin his plans too many times for that. The gardener has learned, perforce, to live with her ambiguities—that she is neither all good nor all bad, that she gives as well as takes away. Nature's apt to pull the rug out from under us at any time, to make a grim joke of our noblest intention. Perhaps this explains why garden writing tends to be comic, rather than lyrical or elegiac in the way that nature writing usually is: the gardener can never quite forget about the rug underfoot, the possibility of the offstage hook.

6. The gardener feels he has a legitimate quarrel with nature—with her weeds and storms and plagues, her rot and death. What's more, that quarrel has produced much of value, not only in his own time here (this garden, these fruits), but over the whole course of Western history. Civilization itself, as Freud and Frazer and many others have observed, is the product of that quarrel. But at the same time, the gardener appreciates that it would probably not be in his interest, or in nature's, to push his side of this argument too hard. Many points of contention that humankind thought it had won—DDT's victory over insects, say, or medicine's conquest of infectious disease—turned out to be Pyrrhic or illusory triumphs. Better to keep the quarrel going, the good gardener reasons, than to reach for outright victory, which is dangerous in the attempt and probably impossible anyway.

7. The gardener doesn't take it for granted that man's impact on nature will always be negative. Perhaps he's observed how his own garden has made this patch of land a better place, even by nature's own standards. His gardening has greatly increased the diversity and abundance of life in this place. Besides the many exotic species of plants he's introduced, the mammal, rodent, and insect populations have burgeoned, and his soil supports a much richer community of microbes than it did before.

Judged strictly by these standards, nature occasionally makes mistakes. The climax forest could certainly be considered one (a place where the number and variety of living things have declined to a crisis point) and evolution teems with others. At the same time, it should be acknowledged that man occasionally creates new ecosystems much richer than the ones they replaced, and not merely on the scale of a garden: think of the tall-grass prairies of the Midwest, England's hedgerow landscape, the countryside of the Ile de France, the patchwork of fields and forests in this part of New England. Most of us would be happy to call such places "nature," but that does not do them (or us) justice; they are really a kind of garden, a second nature.

The gardener doesn't feel that by virtue of the fact that he changes nature he is somehow outside of it. He looks around and sees that human hopes and desires are by now part and parcel of the landscape. The "environment" is not, and has never been, a neutral, fixed backdrop; it is in fact alive, changing all the time in response to innumerable contingencies, one of these being the presence within it of the gardener. And that presence is neither inherently good nor bad.

8. The gardener firmly believes it is possible to make distinctions between kinds and degrees of human intervention in nature. Isn't the difference between the Ile de France and Love Canal, or a pine forest and a condo development, proof enough that the choice isn't really between "all or nothing"? The gardener doesn't doubt that it is possible to discriminate; it is through experience in the garden that he develops this faculty.

Because of his experience, the gardener is not likely to conclude from the fact that some intervention in nature is unavoidable, therefore "anything goes." This is precisely where his skill and interest lie: in determining what does and does not go in a particular place. How much is too much? What suits this land? How can we get what we want here while nature goes about getting what she wants? He has no doubt that good answers to these questions can be found.

9. The good gardener commonly borrows his methods, if not his goals, from nature herself. For though nature doesn't seem to dictate in advance what we can do in a place—we are free, in the same way

evolution is, to try something completely new—in the end she will let us know what does and does not work. She is above all a pragmatist, and so is the successful gardener.

By studying nature's ways and means, the gardener can find answers to the questions, What is apt to work? What avails here? This seems to hold true at many levels of specificity. In one particular patch of my vegetable garden—a low, damp area—I failed with every crop I planted until I stopped to consider what nature grew in a similar area nearby: briars. So I planted raspberries, which are of course a cultivated kind of briar, and they have flourished. A trivial case, but it shows how attentiveness to nature can help us to attune our desires with her ways.

The imitation of nature is of course the principle underlying organic gardening. Organic gardeners have learned to mimic nature's own methods of building fertility in the soil, controlling insect populations and disease, recycling nutrients. But the practices we call "organic" are not themselves "natural," any more than the bird call of a hunter is natural. They are more like man-made analogues of natural processes. But they seem to work. And they at least suggest a way to approach other problems—from a town's decision on what to do with a blown-down pine forest, to society's choice among novel new technologies. In each case, there will be some alternatives that align our needs and desires with nature's ways more closely than others.

It does seem that we do best in nature when we imitate her—when we learn to think like running water, or a carrot, an aphid, a pine forest, or a compost pile. That's probably because nature, after almost four billion years of trial-and-error experience, has wide knowledge of what works in life. Surely we're better off learning how to draw on her experience than trying to repeat it, if only because we don't have that kind of time.

10. If nature is one necessary source of instruction for a garden ethic, culture is the other. Civilization may be part of our problem with respect to nature, but there will be no solution without it. As Wendell Berry has pointed out, it is culture, and certainly not nature, that teaches

us to observe and remember, to learn from our mistakes, to share our experiences, and perhaps most important of all, to restrain ourselves. Nature does not teach its creatures to control their appetites except by the harshest of lessons—epidemics, mass death, extinctions. Nothing would be more natural than for humankind to burden the environment to the extent that it was rendered unfit for human life. Nature in that event would not be the loser, nor would it disturb her laws in the least—operating as it has always done, natural selection would unceremoniously do us in. Should this fate be averted, it will only be because our culture—*our* laws and metaphors, our science and technology, our ongoing conversation about nature and man's place in it—pointed us in the direction of a different future. Nature will not do this for us.

The gardener in nature is that most artificial of creatures, a civilized human being: in control of his appetites, solicitous of nature, self-conscious and responsible, mindful of the past and the future, and at ease with the fundamental ambiguity of his predicament—which is that though he lives in nature, he is no longer strictly *of* nature. Further, he knows that neither his success nor his failure in this place is ordained. Nature is apparently indifferent to his fate, and this leaves him free—indeed, obliges him—to make his own way here as best he can.

What would an ethic based on these ideas—based on the idea of the garden—advise us to do in Cathedral Pines? I don't know enough about the ecology of the place to say with certainty, but I think I have some sense of how we might proceed under its dispensation. We would start out, of course, by consulting "the Genius of the Place." This would tell us, among other things, that Cathedral Pines is not a wilderness, and so probably should not be treated as one. It is a cultural as well as a natural landscape, and to exclude the wishes of the townspeople from our plans for the place would be false. To treat it now as wilderness is to impose an abstract and alien idea on it.

Consulting the genius of the place also means inquiring as to what nature will allow us to do here—what this "locale permits, and what [it] denies," as Virgil wrote in *The Georgics.* We know right off, for instance, that this plot of land can support a magnificent forest of white pines. Nature would not object if we decided to replant the pine forest. Indeed, this would be a perfectly reasonable, environmentally sound thing to do.

If we chose to go this route, we would be undertaking a fairly simple act of what is called "ecological restoration." This relatively new school of environmentalism has its roots in Aldo Leopold's pioneering efforts to re-create a tall-grass prairie on the grounds of the University of Wisconsin Arboretum in the 1930s. Leopold and his followers (who continue to maintain the restored prairie today) believed that it is not always enough to conserve the land—that sometimes it is desirable, and possible, for man to intervene in nature in order to improve it. Specifically, man should intervene to re-create damaged ecosystems: polluted rivers, clear-cut forests, vanished prairies, dead lakes. The restorationists also believe, and in this they remind me of the green thumb, that the best way to learn about nature's ways is by trying to imitate them. (In fact much of what we know about the role of fire in creating and sustaining prairies comes from their efforts.) But the most important contribution of the restorationists has been to set forth a positive, active role for man in nature—in their conception, as equal parts gardener and healer. It seems to me that the idea of ecological restoration is consistent with a garden ethic, and perhaps with the Hippocratic Oath as well.

From the work of the ecological restorationists, we now know that it is possible to skip and manipulate the stages of forest succession. They would probably advise us to burn the fallen timber—an act that, though not strictly speaking "natural," would serve as an effective analogue of the natural process by which a forest is regenerated. The fires we set would reinvigorate the soil (thereby enhancing *that* wilderness) and at the same time clear out the weed species, hardwood saplings, and brush. By doing all this, we will have imitated the conditions under which a white pine forest is born, and the pines might then return on their own. Or else—it makes little difference—we could plant them. At that point, our work would be done, and the pine forest could take care of itself. It would take many decades, but restoring the Cathedral Pines would strain neither our capabilities nor nature's sufferance. And in doing so, we would also be restoring the congenial relationship between man and nature that prevailed in this place before the storm and the subsequent controversy. That would be no small thing.

Nature would not preclude more novel solutions for Cathedral Pines—other kinds of forest-gardens or even parks could probably flourish on this site. But since the town has traditionally regarded Cathedral Pines as a kind of local institution, one steeped in shared memories and historical significance, I would argue that the genius of the place rules out doing anything unprecedented here. The past is our best guide in this particular case, and not only on questions of ecology.

But replanting the pine forest is not the only good option for Cathedral Pines. There is another forest we might want to restore on this site, one that is also in keeping with its history and its meaning to the town.

Before the storm, we used to come to Cathedral Pines and imagine that this was how the New World forest looked to the first settlers. We now know that the precolonial forest probably looked somewhat different—for one thing, it was not exclusively pine. But it's conceivable that we could restore Cathedral Pines to something closely resembling its actual precolonial condition. By analyzing historical accounts, the rings of fallen trees, and fossilized pollen grains buried in the soil, we could reconstruct the variety and composition of species that flourished here in 1739, the year when the colonists first settled near this place and formed the town of Cornwall. We know that nature, having done so once before, would probably permit us to have such a forest here. And, using some of the more advanced techniques of ecological restoration, it is probably within our competence to re-create a precolonial forest on this site.

We would do this not because we'd decided to be faithful to the "state of nature" at Cathedral Pines, but very simply because the precolonial forest happens to mean a great deal to us. It is a touchstone in the history of this town, not to mention this nation. A walk in a restored version of the precolonial forest might recall us to our culture's first, fateful impressions of America, to our thoughts on coming upon what Fitzgerald called the "fresh green breast of the new world." In the contemplation of that scene we might be moved to reconsider what happened next—to us, to the Indians who once hunted here, to nature in this corner of America.

This is pretty much what I would have stood up and said if we'd had a town meeting to decide what to do in Cathedral Pines. Certainly a town meeting would have been a fitting way to decide the matter, nicely in keeping with the genius of *this* place, a small town in New England. I can easily imagine the speeches and the arguments. The people from the Nature Conservancy would have made their plea for leaving the place alone, for "letting nature take her course." Richard Dakin, the first selectman, and John Calhoun, the forest's nearest neighbor, would have warned about the dangers of fire. And then we might have heard some other points of view. I would have tried to make a pitch for restoration, talking about some of the ways we might "garden" the site. I can imagine Ian Ingersoll, a gifted cabinetmaker in town, speaking with feeling about the waste of such rare timbers, and the prospect of sitting down to a Thanksgiving dinner at a table in which you could see rings formed at the time of the American Revolution. Maybe somebody else would have talked about how much she missed her Sunday afternoon walks in the forest, and how very sad the place looked now. A scientist from the Yale School of Forestry might have patiently tried to explain, as indeed one Yale scientist did in the press, why "It's just as pretty to me now as it was then."

This is the same fellow who said, "If you're going to clean it up, you might as well put up condos." I can't imagine anyone actually proposing that, or any other kind of development in Cathedral Pines. But if someone did, he would probably get shouted down. Because we have too much respect for this place; and besides, our sympathies and interests are a lot more complicated than the economists or environmentalists always seem to think. Sooner than a developer, we'd be likely to hear from somebody speaking on behalf of the forest's fauna—the species who have lost out in the storm (like the owls), but also the ones for whom doing nothing would be a boon (the beetles). And so the various interests of the animals would be taken into account, too; indeed, I expect that "nature"—all *those* different (and contradictory) points of view—would be well represented at this town meeting. Perhaps it is naïve of me to think so, but I'm confident that in the course of a public, democratic conversation about the disposition of Cathedral Pines, we would eventually arrive at a solution that would have at once pleased us and not offended nature.

But unfortunately that's not what happened. The future of Cathedral Pines was decided in a closed-door meeting at the Nature Conservancy in September, after a series of negotiations with the selectmen and the owners of adjacent property. The result was a compromise that seems to have pleased no one. The fallen trees will remain untouched—except for a fifty-foot swath clear-cut around the perimeter of the forest, a firebreak intended to appease the owners of a few nearby houses. The sole human interest taken into account in the decision was the worry about fire.

I drove up there one day in late fall to have a look around, to see what the truce between the Conservancy and the town had wrought. What a sad sight it is. Unwittingly, and in spite of the good intentions on both sides, the Conservancy and the selectmen have conspired to create a landscape that is a perfect symbol of our perverted relation to nature. The firebreak looks like nothing so much as a no-man's-land in a war zone, a forbidding expanse of blistered ground impounding what little remains of the Cathedral Pines. The landscape we've made here is grotesque. And yet it is the logical outcome of a confrontation between, on the one side, an abstract and mistaken concept of nature's interests and, on the other, a pinched and demeaning notion of our own interests. We should probably not be surprised that the result of such a confrontation is not a wilderness, or a garden, but a DMZ.

Winter

CHAPTER 11

"Made Wild by Pompous Catalogs"

Winter in the garden is the season of speculation, a time when the snow on the ground is an empty canvas that invites the idle planting and replanting of countless hypothetical gardens between now and spring thaw. A season of speculation in the Wall Street sense too, for now is when large wagers of gardening time and space are made on the basis of mere scraps of information—a hankering, a picture in a catalog, a seed. We gardeners have always had trouble heeding Henry Ward Beecher's sound nineteenth-century advice, that we not be "made wild by pompous catalogs from florists and seedsmen."

In a few months, summer will pass judgment on the merit, or folly, of our January schemes, but right now anything seems possible. The winter garden is an abstract, earthless place where only representations bloom—the season's lists and sketches and catalogs and seeds (which are of course nature's own representations). Insubstantial as these seem, they are in fact as vital to the summer garden as water and humus and sunlight. For it's the gardener's traffic in such signifiers that, not unlike the traffic of bumblebees in summer, rejuvenates the garden, importing the fresh genes and novel combinations that each year make it new.

The bumblebee consults his blossoms and the gardener his catalogs, which blossom extravagantly at this season, luring him with their four-color fantasies of bloom and abundance. Catalogs lie at the center of the winter garden. Through their pages the gardener, who has worked in isolation all summer, steps out into the wider gardening world and returns with a rich store of new information—genetic, horticultural, and cultural. The genetic information comes in the form of the seeds offered for sale, of course, and the horticultural information in the form of the valuable advice many catalogs contain. (I have learned much about perennials from White Flower Farm's patient, didactic text; from Seeds Blum I've learned the ins and outs of saving and starting my own seeds.) For the cultural information the gardener has to read between the lines, but there it is: many of the seed and plant catalogs are subtle (and sometimes not-so-subtle) compendia of social, political, and moral instruction.

After you've read a dozen or so catalogs, you start to realize that the differences among them are not so much in the varieties of plants and seeds they offer, for there is a great deal of overlap, as in the distinctive way each of them chooses to imagine the garden. Among other things, a garden is a form of self-expression, and we page through the various catalogs looking for the elements of a vocabulary that suits us, that can give body to our wishes. From White Flower Farm or Wayside Gardens we can have a perennial border that fairly bristles with class distinctions, floral testimony to our sophistication; from Harris or Park or Gurney's we can order a middle-class garden that proudly announces to the neighbors our family's enterprise, independence, and togetherness; from Johnny's Selected Seeds or Pinetree Garden Seeds we can get a garden that reflects our environmental consciousness; and from Seeds Blum or J. L. Hudson one that proclaims our political convictions, in particular our zeal to protect the planet's genetic diversity from the depredations of big business. There are many versions of the garden, each with its far-flung subscribers and a catalog or two to forge them into a community that, like most communities, tends to define itself in opposition. Spend a few quiet winter nights with these not-so-quiet catalogs, and you begin to see that, just beneath its placid surface, the garden is buzzing with social and political controversy.

Standing at the top of the gardening world's social hierarchy, like two great aristocratic families, are the catalogs of White Flower Farm, in Litchfield, Connecticut, and Wayside Gardens, in Hodges, South Carolina. Both are printed in full color on coated stock, and both offer a matchless selection of high-quality perennials and shrubs, backed up by generous guarantees and the kind of impeccable service one expects at these prices, which can be dizzyingly steep. But there the similarities end, for Wayside and White Flower, though they sell many of the same plants, nevertheless propose entirely different conceptions of higher gardening, about as different as you'd expect the styles of great Southern and Yankee families to be.

Amos Pettingill, the fictional character who signs the White Flower catalog (which doesn't even call itself a "catalog"—that would be too mercenary—but *The Garden Book*) sets the tone for this house: amiably eccentric, opinionated, prudent, aloof from commerce, afflicted with a bad case of anglophilia, ironically self-deprecating and yet at the same time (a trick only the well bred seem able to pull off) coercive on matters of taste. I should say right here that I can't stand Amos Pettingill. I hate his prissy, Edwardian prose, the way he speaks of a climbing rose as a "frightful misnomer" or of the weather in zone one as "so beastly cold." I hate when he tries to flatter me by referring to my yard as "your grounds." And I hate all his *entre-nous* asides: "It is possible, of course, for a good gardener to be impatient, but we have noted over the years that impatient gardeners all have money. They buy time in the form of specimen plants and build a garden overnight, and probably get as much satisfaction out of building quickly [it's right here that the polygraph needle leaps] as those of us who build more with time than money." Here is the classic distinction between old wealth and new, narrowly rescued from the charge of snobbery by Pettingill's kind, condescending fib. It is of course off the proceeds of impatient gardeners, those untutored nouveaux riches whom the catalog takes such pains to instruct, that White

Flower Farm has—rather quickly—built itself into an $8 million business.

But I think I hate most of all the invitation, included in each spring catalog, to join Pettingill and the staff at White Flower Farm's annual Open House "for cucumber sandwiches and iced tea on the lawn in front of our house." This summer the big event takes place on July 14, Bastille Day. I would like nothing better than to be able to pass my invitation on to Abbie Hoffman, or some other deserving *sans-culotte.* What I would give to watch John Belushi work out on a stack of those finger sandwiches, or to hear Hunter Thompson demand something a little stiffer to drink than iced tea. Get me the Marx Brothers! Can you imagine the kind of people who would actually turn out for such an affair? I have a nightmare vision of a broad lawn upon which whale pants from Brooks and wrap skirts from Talbots come and go, talking of Gertrude Jekyll, or the *beastly* heat here in zone five.

You should probably also know that none of this has stopped me from shopping at White Flower Farm. Most of what I know about the culture of perennials I learned at the feet of Amos Pettingill. I trust him implicitly when he tells me that Fama is "the finest cultivar of the perennial scabious," or that, although "the lax habit of *Veronica latifolia* calls for some stakes or twiggy supports to look its best, it is well worth the effort. Plant several beside white oriental poppies and back them up with *caryopteris.*" Perhaps because I garden nearby, in the same climate and in the same geographic and architectural (if not social) context, I find he is almost always right. Huffy and reactionary as it often is, his judgment is heeded. White Flower is notoriously slow to introduce new cultivars (and annoyingly boastful about it: "In a world addicted to change for its own sake . . ." etc.), but when they finally do endorse a coreopsis Moonbeam, or a Stella D'Oro day lily, you can plant it with confidence. No one gets into Pettingill's club before being thoroughly vetted. "New and improved" is a term of opprobrium in White Flower's catalog, a sarcastic jibe at the garden world's racier, hypier outfits. Here a rediscovered old rose will always stand a better shot at admittance than a shiny new hybrid.

A certain snobbery about plant genes is often justified—how many shiny new hybrids ever live up to their billing?—but sometimes it is really just a mask for a much less benign snobbishness about human bloodlines and society. Take the issue of color, a subject in the garden that almost seems to vibrate with questions of status. Why *White* Flower Farm? Because, in the words of the firm's founder (cribbed from Vita Sackville-West), "White flowers are anathema to all but the oldest and most sophisticated gardeners." In the hierarchy of flower colors, white is unquestionably the highest—so much so, says Eleanor Perényi (always a reliable arbiter of taste in the garden), that "even a white gladiolus may get by—just—while one in any other color is beyond the pale." (But wait, what's this about gladioluses? Well, it seems they're . . . problematic, as a careful reading of the White Flower catalog discloses: "Gladioli must be used carefully in the border," Pettingill cautions, "for most of them are rather stiff . . . and gaudy.")

The spectrum of plant snobbery—from white to blue to pink to yellow, red, orange, and finally magenta, the too-common banner of nature's proletariat, the weeds—is implicit on every page of the White Flower Farm catalog. Whites and blues predominate, and although flowers in the hotter hues do make appearances, these are doled out judiciously, always in the subtlest tones available and frequently accompanied by a warning to the wise from the gardener-general. The perennial border conjured by the White Flower catalog is a restrained and subtle affair, befitting its New England provenance. (The rules relax only in autumn, when, perforce in New England, subtlety goes out the window; in September a White Flower garden will make room for even the showiest dahlias—"breathtaking competition for the early fall foliage.") White Flower's summer border is one that's meant to work well with old stone walls and clapboard; if it errs, it will be on the side of austerity, never extravagance. Here is an aesthetic to please even a Puritan.

If White Flower proposes a garden fit for Cabots and Lodges, Wayside has one Scarlett O'Hara would die for. From the photographs alone it's possible to distinguish Amos Pettingill's subdued Connecticut aesthetic from the more demonstrative one operating in Hodges, South Carolina. Wayside likes to photograph its flowers when they are in full bloom, wide open and almost past their peak—as in one of those Dutch still lifes where blossoms poise on the verge of shattering and decadence hovers in the wings. Wayside blooms press forward from their pages provocatively, many of them bursting free of their frames, almost as if from a bodice. The effect is frankly sensual, yet never quite garish—except, perhaps, to a Puritan eye.

I imagine the voluptuousness of the Wayside catalog must scandalize the Yankee competition up in Litchfield, where flowers are always photographed at a discreet distance and several days before their peak. Set before me are a Wayside and a White Flower photograph of the same flower, a Madame Hardy rose. In White Flower's picture, she appears somewhat chaste, her innumerable white petals not yet open all the way; something is withheld. In Wayside, the same flower is shot several days later, her petals now fully opened to expose the green button-eye stamen within. Wayside's catalog copy extols her "slight blush," "delicious fragrance," and blooms "produced freely in June." White Flower refuses to effuse, reporting matter-of-factly that "Mme. Hardy has earned a spot at the back of the Moon Garden adjoining our house, where she is most welcome." Wayside revels in the sexiness of its flowers, acknowledging their claims on our senses, in a way that White Flower would not dare. White Flower likes to think of its flowers as dowager aunts or dutiful daughters, but never as sex objects. If Amos Pettingill had to think about flowers in *that* way, he would probably pop his buttons.

Yet for all its sensationalism, the Wayside catalog never lets anyone forget its upper-class connections, a knack that strikes me as peculiarly southern. Confident of its status and uninhibited by Yankee Calvinism, the southern garden seems able to indulge the hotter, more emotive flowers and colors without the slightest hint of embarrassment. Unabashed, Wayside carries plants whose very names would make Amos Pettingill squirm with self-consciousness: roses named Cupcake, Rise 'n' Shine, and Gina Lollobrigida, a candy-colored lily named Strawberry Shortcake. And Wayside's catalog is as free with its printer's supply of magenta as White Flower is with cyan. That's because, according to the rules of this particular world, a flower can be showy, cute, or coquettish without dishonoring the family name. She can get away with it—because of who her daddy is. And indeed Wayside catalog copy devotes an inordinate amount of space to the subject of parentage. All of its plants are "pedigreed," the catalog informs us, because "you want plants of superior horticultural refinement and purity. Breeding in plants, just as in animals, is the only way to assure pure, true strains and prize specimens."

If status in the White Flower world is a matter of discrimination and taste, in Wayside it is one of ancestry and honors. Wayside inevitably tells us who the parents of its roses are, and duly notes the date of their debut in society. Last year's featured rhododendron, a frilly pink concoction as overdone as an antebellum ball gown, warrants our attention because it was bred by none other than Fred Peste, the "highly respected plantsman" of Washington State, where "there is a long tradition of fine rhododendron breeding." This Wayside exclusive "includes in its parentage *R. Yakushimanum,* one of the most influential rhododendrons in modern breeding—and from which it gets its compact habit, heavy flowering, and stunning colors." No surprise, then, that, "in the face of very stiff competition," this Wayside exclusive was "selected to be honored in a special ceremony in Olympia in 1989." Who cares what she looks like? I'll marry her!

Not too surprisingly, any plant boasting the slimmest connection to European royalty will win Wayside's heart. Rose Princesse de Monaco, a bicolored hybrid in ivory and "rouge" that is "now proving an exceptionally fine rose for the show table" (that's catalog code for "It's a disaster in the garden"), is another Wayside exclusive. Get out your handkerchiefs:

> The tragic passing of Princess Grace of Monaco emphasized the touching personal history of this rose. Since the 1956 wedding of Prince Rainier and Grace Kelly, there developed a close friendship between the royal family and the House of Meilland, for years one of the world's leading rose hybridizers. . . . In 1973, Meilland dedicated a new rose to the young Princess Stephanie, and in 1976 a rose garden was dedicated to Francis Meilland by Monaco, which in 1981 hosted the first International Rose Show.
>
> Among the new roses that the Meillands exhibited at that show was a bicolor with rouge and cream buds that opened into perfect, full blooms of rich ivory, tinged with a deep pink blush (the colors of the Principality). . . . Initially exhibited under the name 'Preference,' its credentials were enhanced by having the world famous 'Peace' as an ancestor. Princess Grace immediately declared it her favorite, and Alain Meilland decided to rename it 'Princesse de Monaco.'

Wayside probably lays so much more stress on pedigree than does White Flower (Amos would have decided quite rightly that Princess Grace had no taste in roses and left it at that) because the firm is more heavily invested in newer hybrids, which is the touchiest of subjects in the garden world. In general, a gardener's tolerance of modern hybrids is a mark of middle-brow taste. Modern hybrids are the garden world's parvenus, the offspring of dubious unions that threaten to dilute the bloodlines of plant society. Wayside's aesthetic happens to favor the kind of intense color and frilliness that hybridization makes possible, but the firm overcomes the potential stigma by endorsing only the "best crosses," the offspring of carefully arranged marriages between the most distinguished of families.

The big mainstream seed catalogs—Burpee, Park, Harris, Stokes, Gurney's—have no such compunctions about hybrids. In fact they love nothing better than a novel cross, the more improbable the better. Bigger, better, newer, just plain *differenter*—these are the supreme values of what I think of as the middle-class catalogs. Of course the American middle class is a house with many floors, and it's necessary here to make some distinctions: Burpee is relatively upscale these days, offering even a few European vegetables and old roses; at the other end of the spectrum stands Gurney's, a downscale tabloid of a catalog laid out and edited in the spirit of the *National Enquirer.* Park, Harris, and Stokes are somewhere in the middle. What joins all these catalogs is their worship of the new; for better or worse, they represent the triumph of progress and middle-class taste in the garden.

At the beginning of each of these catalogs you will invariably find an overinked and overhyped section jam-packed with new offerings, the proud (and often patented) creations of the house's plant breeders and geneticists. The efforts of these lab-coated wizards seem to be channeled in one of two directions: toward ridding a popular plant of an allegedly debilitating trait, or toward endowing it with some fantastic new feature. In the former category are zucchini with smooth stems ("Look Ma!" the Park catalog exults, "no scratches!"), "burpless cucumbers" (yet another world problem cracked by American know-how), and watermelons without seeds.

But it is the world-beating, all-new, exclusive features these companies dream up each year that get the most ink in their catalogs—and draw the heaviest fire from garden-world traditionalists. Sometimes the industry's innovations make immediate sense: cantaloupes that can grow in Alaska, day lilies that bloom all summer, self-blanching cauliflowers, disease-resistant tomatoes. Whether or not these hybrids live up to their billing, one can appreciate the rationale behind them. But many new hybrids are somewhat less intelligible. Since at least the 1950s, the seedsmen have been hard at work creating zinnias that look like dahlias, dahlias that look like zinnias, chrysanthemums that look like daisies or cacti or zinnias, zinnias that look like chrysanthemums, and all of these in previously unheard-of hues. "Now I *like* chrysanthemums," Katherine White wrote with a touch of exasperation in 1958, "but why should zinnias be made to look like them?"

For many years now Burpee has had a peculiar obsession with developing a marigold that would have no gold in it. Beginning in the fifties, the firm offered ten thousand dollars to anyone who happened upon a white marigold—what is called a "sport," or chance mutation. The prize was finally claimed by some lucky gardener in 1975, and in this year's catalog, Burpee proudly offers, on an exclusive basis, its Snowdrift marigold, "the culmination of thirty years of white marigold research." Looking at the photograph, one has to wonder if the effort was worth the candle: Snowdrift marigold looks exactly like a white chrysanthemum.

But surely the all-time most pointless product of the hybridizer's science is Park's latest triumph, the Sunspot sunflower. The catalog's photograph of this "stunning, delightful" breakthrough shows a toddler who is *literally towering over a field of sunflowers.* For reasons I cannot begin to fathom, Park has put full-sized sunflower blossoms on top of two-foot stalks. The result is an awkward-looking pygmy, a floral geek that could only have been designed by people who never stopped to consider what a sunflower is—which is, first and foremost, and last and in-between, *tall.*

Should Amos Pettingill ever deign to carry mere annuals and vegetables in his catalog, you can be sure he will pass over the Sunspot sunflower and Snowdrift marigold. On grounds of aesthetics, I am with him. But the upper crust of the garden world voices its objections to such hybrids with a vehemence that makes you wonder if perhaps something more than good taste is at stake here. The hybridists, Eleanor Perényi grumbles, "have introduced previously unheard-of vulgarisms to the garden world"; she avows her "contempt for the grosser hybrids." Like Katherine White and Amos Pettingill, she deplores any flower made to behave in an "unnatural" manner: "chartreuse narcissus, mauve day lilies, pink forget-me-nots and all those bi-colors inflicted wherever possible."

Gardening's WASPish old guard simply can't abide the horticultural melting pot bubbling over in the pages of the middle-brow catalogs—all these questionable marriages among flowers who've forgotten their proper place in the world. The nerve of a dahlia to start acting as if it were a zinnia or chrysanthemum! And what is that common day lily doing all dressed up in mauve? That color is the private property of the lilacs! Let us have none of this reckless gene-swapping or we shall soon have no more "pure, true strains." Do I detect perhaps a touch of anxiety about miscegenation here?

Heedless of these guardians of genetic purity, the middle-class catalogs go blithely on parading their annual improvements with an enthusiasm it's hard not to like. Progress is their religion, and if it sometimes serves up a dud like the Sunspot sunflower, it has also given us many good things, like the sugar snap pea. This edible-podded pea became one of the most popular vegetable introductions of the eighties, Eleanor Perényi's huffing ("second-rate" and "inedible") notwithstanding. The sugar snap is the Teflon pan or microwave oven of the garden world, much quicker to win converts in the middle-class than among food snobs. But eventually the nouvelle cuisine came around and endorsed the sugar snap, and now you'll find this upwardly-mobile pea even in high-brow vegetable seed catalogs such as Cook's and Shepherd's, firms that cater to serious cooks.

At their best, the middle-brow catalogs capture something of the boundless optimism of the American Century, which can seem preserved in their pages as if in a time capsule. Here, still going strong, you'll find ruddy specimens of the nuclear family, clean-cut kids who look like they'd much rather help out Mom and Dad with the yard work than play video games at the mall or experiment with drugs. Here, too, the promise of American technology still shines brightly, nowhere more so than in the all-important realm of tomato development. The state-of-the-art hybrids all carry streamlined, space-age names like Supersonic, Ultra Boy, Jetstar, Starshot, and the radioactive-sounding Fireball. Stokes' latest catalog is puffed up with pride over the news that NASA has selected its new Lunch Box tomato "as a possible variety to feed the new space station workers." This might seem a dubious endorsement (these are, after all, the people who brought us Tang and freeze-dried beef stroganoff), but I have to assume Stokes knows its clientele better than I do.

My favorite throwback catalog is Gurney's of Yankton, South Dakota, a cheerfully low-brow tabloid printed in saturated colors on cheap, uncoated stock. Nothing seems to have changed here since before the war, not even the prices (bleeding heart plants: thirty-nine cents with any order). To page through Gurney's is to revisit a simpler, perhaps more ingenuous America, a country of the can-do where the values of thrift, independence, and family have lost none of their sheen. I picture Gurney's spread out on farmhouse kitchen tables in the thirties and forties, maybe even in Dorothy's Kansas. Gurney's customers plant gardens not to make a social statement, but to economize, to spruce up the place a bit ("flowers," the catalog says, "are like frosting" on a house), and to keep the kids out of trouble. "For the last two years," reads one of the numerous testimonials reprinted in the catalog, "my sons and I have won first-place ribbons in the Tennessee State Fair using Gurney's Cobb Gem watermelon seeds." In the pages of Gurney's you can still find the makings of a Victory Garden; indeed, here are all of the dusty, antique accoutrements of the Agrarian Ideal.

Gurney's seems to be a world where hamburgers are served regularly for dinner, perhaps with corn on the cob. This year the catalog is very high on its Walla-Wallas, the "burger-sized onion" pictured on the cover, where a trio of these golden beauties are shown tipping a scale at more than four pounds. "Extra-large slices easily cover the bun—even on quarter-pound burgers. No more bites of just bread—you get onion every time!" (True luxury, to consider this a problem.) In fact most of the catalog's onion entries make reference to hamburgers, evidently the raison d'être of any self-respecting onion. Though Gurney's does dish up a few high-class Sunday dinners, too: a plate of its Jersey asparagus is shown piled high beneath a ribbon of hollandaise (and, inexplicably, accompanied by several pieces of gold-plated tea service). The catalog boasts that its Tom Thumb lettuce is actually served at New York City's

Waldorf-Astoria Hotel, a place that, from out there in Yankton, South Dakota, shimmers in the distance, an El Dorado of elegance.

As you might expect, Gurney's prose style leans toward the hyperbolic; the key literary influence here would appear to be Ripley's Believe It or Not, and exclamation points sprout from the page like weeds after a rain. Gurney's is proud to offer the World's Largest Watermelon (Cobb Gem, "a mammoth melon that's sure to make your neighbor's eyes bulge!"), the World's Largest Pumpkin (Atlantic Giant, reaching up to a whopping 755 pounds), the World's Largest Tomato (Delicious, current holder of the Guinness Record), and even the World's Largest Radish (German Giants, which grow "as big as baseballs!").

Bigger *is* better, and to drive home just how much better Gurney's vegetables are, frequently the photographs will include some handy visual reference to help establish the scale: a dime, a coffee mug, a dinner plate heaped with corncobs, a beaming toddler peeking over the horizon of a mammoth pumpkin. Among other things, gardening in Gurney's world is a form of entertainment, and gee-whiz novelty items abound: the white tomato, the leafless pea ("a pea picker's paradise!"), the blue rose, the five-in-one apple tree (five different varieties grafted onto a single tree), and that old exotic standby, the Venus flytrap.

I have to admit that, as much as I enjoy this catalog, I do more browsing than shopping in Gurney's. Most of its offerings sound too good (and too cheap) to be true. Maybe it's because Gurney's world is by now so alien and antiquated, but I find it almost hard to believe that these seeds would germinate in the present, in the America I live in. Gurney's is the ghost of gardening past.

This is nowhere more plain than in its section on pest controls, which has a definite pre–*Silent Spring* slant to it. The miracles of modern chemistry have lost none of their glamour in the pages of Gurney's, whose long line of petrochemical wonders includes Corry's Slug and Snail Death, big bags of an unspecified insecticide blend called simply Bug Dust, and the truly alarming CPF, an insecticide that is mixed into house paint in order to turn your "home, garage, or outbuildings into giant bug zappers." This sort of uncritical faith in technology is fading from the gardening world (indeed, it began losing adherents here long before anywhere else), and the more forward-looking of the mainstream houses (Burpee, for one) have begun to back off chemicals and to stress more organic approaches. If it hasn't done so already, history may soon leave Gurney's (along with several other of the more traditional middle-class catalogs) behind, in a cloud of Bug Dust.

Emerging in the last few years to take their place are a small but flourishing group of "counterculture" catalogs that define themselves in opposition to the big mainstream seed houses. With perhaps just a whisper of righteousness, the Vermont Bean Seed Company, Johnny's Selected Seeds, and Pinetree Garden Seeds (the last two from Maine) stress organic practices, shy away from novel hybrids, and studiously avoid any boastful world-beating rhetoric. These catalogs have a distinct rural flavor but, unlike Gurney's, here it's once removed, filtered through a disillusionment with the urban experience. I imagine these catalogs being written by ex-hippies who went back to the land in the seventies and stayed on. Here are all the rural virtues, but deliberately achieved now and held up as a mild rebuke of modern ways. "We never forget," Rob Johnston of Johnny's intones at the beginning of one recent catalog, "that in addition to the need for productivity, the food will be eaten, and should be enjoyable and nourishing. This spirit fuels our work."

"Your orders make it possible for us to do the interesting and enjoyable work we do," Dick Meiners, the proprietor of Pinetree, writes. "You, our customers, are continually in our thoughts, both as individuals and as a group." Doesn't this sound like the sort of company Jimmy and Rosalynn Carter might order their seeds from? Everything about these catalogs strikes me as very much a product of the seventies, when in fact most of them were started: they're modest and sincere, wary of big corporations and technology, environmentally scrupulous, and resolutely sober. In the pages of these catalogs it feels less like the American Century than the morning after.

Distinctly more up-to-date and cosmopolitan than Johnny's or Pinetree are the catalogs of Cook's, in Vermont, and Shepherd's, in Northern California. These two seed boutiques specialize in imported vegetables, on the premise that the hybrid varieties offered in the mainstream catalogs are as bland and homogenized as the supermarket's. In search of culinary novelties, Cook's and Shepherd's look not to the laboratory but to Italy, France, or Japan, countries which seem to have outstripped us in the quality of their produce as well as their technology. These houses aim to rescue the flavor and wholesomeness of vegetables from the sins of technology and mass production; they are to Gurney's and Stokes what Alice Waters is to Ray Kroc. (And so are their prices.) Here, iceberg lettuce is as outré as a gladiolus. What you will find instead are *haricots verts,* radicchio, mesclun, arugula, mâche, and, in Cook's, no fewer than nine varieties of basil. But not a single burger onion.

All five of these catalogs put forward a critique of the hybrid, but it's not the same critique lodged by the garden world's upper crust. Though considerations of status obviously carry weight in Cook's and Shepherd's, in general the countercatalog's brief against the hybrid turns less on social questions than on vaguely moral and political ones—it comes more from the left than the right. The premise of these relatively new businesses is that the mass-market seed companies have forgotten the little guy. Most of the hybrids on the market have been bred to meet the needs of commercial growers, who care much less about a vegetable's flavor and healthfulness than about its ability to withstand mechanical harvesting and cross-country shipment. Order from Cook's or Johnny's and you strike a blow for an older, more wholesome way of doing things. These companies occupy the same (evidently expanding) crack in the national economy that Ben & Jerry's, independent bookstores, and L. L. Bean do.

If patronage of this kind of business offers a faint moral tingle, a much more potent political jolt may be had from Seeds Blum and J. L.

Hudson, the fiery radicals of the garden world and, this winter, the two catalogs I seem to have spent the most time with. Here in their pages the battle against the hybrids and their corporate sponsors rises to the level of political struggle, if not moral crusade.

What is at stake in this struggle is best gleaned from the patient, accessible (even somewhat boppy) text of Seeds Blum. This catalog looks exactly like my high school literary magazine—unmistakably homemade, with the same 8½-by-11 format and matte stock, and the same sweet hippy-dippy drawings. (Imagine the California Raisins as drawn by R. Crumb.) Seeds Blum, which was founded in Boise, Idaho, in 1982 by Jan Blum, specializes in what are called heirloom vegetables, the seeds of old varieties that have been lost to commerce in the rush to create and market new F-1 hybrids. The seeds of an F-1 hybrid, Blum explains, are those that come from the first generation of a new cross. The plants produced by these seeds will be genetically identical, but *their* seeds will not "come true"—either they will be sterile or the less-desirable traits of one or the other parent will express themselves in the second generation.

These characteristics make the F-1 hybrid the road on which advanced capitalism invades the garden and farm. A field of genetically identical corn or tomatoes is not only perfectly uniform in taste and appearance (thereby answering to mass-market demands), but it will ripen all at once, which allows the farmer to harvest efficiently by machine. By helping to bring the techniques of mass production to agriculture, the F-1 hybrid makes the modern factory farm possible. For the seed company, the advantages of hybrids are even greater: as the Seeds Blum catalog puts it, in words as plain as Marx and Engels's, "the reason hybrids exist is to protect the breeding investment of the seed company." A new F-1 hybrid can be patented—it is a form of private property. And since the seeds produced by such a hybrid are themselves worthless, the companies holding the patents have succeeded in making the farmer and the gardener their dependents; the companies have gained control of the "means of production." Thus does the F-1 hybrid remake nature in the image of capitalism.

Why is this necessarily a bad thing? Jan Blum's objections seem compelling: besides being wearisome for the consumer and the gardener, the homogeneity of hybrid crops is dangerous, for the reason that genetic uniformity invites epidemics. Once one plant in such a totalitarian crop succumbs to a disease or insect, the odds are strong that all the others will, too. Today, American farmers plant only a handful of different hybrid varieties of corn and, as they discovered in 1970, a single blight can now decimate the nation's entire crop. By narrowing the genetic base of our agriculture we have made it much more vulnerable and, in turn, more dependent on chemical defenses. (It is no coincidence that several of the big seed houses are now owned by chemical companies.) The same uniformity that smooths capitalism's way also contradicts one of nature's cardinal principles, which is genetic diversity.

By preserving and disseminating heirloom seeds, which are "open pollinated" (that is, they can reproduce themselves in nature—though like any living thing that reproduces sexually, never quite the same way twice), we help to keep the gene pool wide and deep. Many of these varieties have been kept alive over the years by individual gardeners, who selected and saved the seeds of particular plants that possessed traits they prized: fitness for particular local condition (something about which the national seed companies couldn't care less), resistance to disease, and, most important of all, flavor. By passing down seeds from generation to generation, these gardeners have functioned literally as a force of natural selection, evolving scores of excellent varieties and countless traits upon which the survival of our agriculture might someday depend. Today, their cumulative work constitutes a trove of genetic information that, if not for the work of Jan Blum and others like her, might otherwise soon be lost.

Blum's catalog, equal parts sermon and horticultural be-in, enlists the gardener in the mission of keeping these species alive. Here I can buy the seeds of the native-American Sibley squash, Purple Calabash tomatoes, and Jenny Lind, the prized muskmelon of the nineteenth century. I can also find out from the Seeds Blum catalog how to start seeds, save my own seeds next fall, and even how to go about trading them with other gardeners. For among other things, Seeds Blum functions as a kind of seed-saver's co-op, encouraging gardeners to share the fruits of their labor through her "Trading Post," even though this service must cut down on the seed Jan Blum can sell. Seeds Blum is the garden world's equivalent of the Grateful Dead, who, at no small financial sacrifice, encourage their fans to trade concert tapes among themselves. It's not the money but the plants, man—that and Jan Blum's vision of a time when the "seeds once again will belong to the people rather than the seed companies."

If Jan Blum is a sixties radical come to the garden, J. L. Hudson reads like an ornery, long-in-the-tooth thirties anarchist, and in fact Emma Goldman is quoted extensively in this stern, somewhat forbidding catalog. At nearly a hundred pages, crammed margin-to-margin with six-point type, *The Ethnobotanical Catalog of Seeds* is an extraordinary document. In addition to the seeds of more than one thousand different plants (a great many of them available nowhere else), the Angry Man of Redwood City, California, lards his catalog with quotations from his favorite philosophers (who range from Goldman to Bob Dylan and William Burroughs to Lydia Hyde Bailey and Lao-Tzu), his own sharp-tongued and enlightening tracts on such subjects as "Inter-Species DNA Transfer" and "The Value of Human Diversity," and an idiosyncratic selection of books, not all of them directly about seeds or gardening. In addition to *The Useful Native Plants of Australia* and *Ethnobotany of the Hopi* you will find this: "*U.S. Constitution.* 1 oz.; $1. Everyone should have a copy of the Constitution in their home. It is startling to read the Bill of Rights (first ten amendments) in view of how few are still respected by government. Read it and weep."

J. L. Hudson's vision of the garden is completely original. It begins with the kind of radical critique of hybrids that Jan Blum makes, but then it broadens to take in the meaning of evolution and the proper role of man on the planet. As best as I can piece his philosophy together from the catalog text, Hudson espouses a distinctive brand of libertarian politics-cum-genetics. In his view, the highest calling of *Homo sapiens* is as a kind of higher-order bumblebee or hummingbird, charged with the mission of disseminating plant genes across the face of the Earth in order to advance the work of evolution. Anyone standing in the way of this work—big seed houses, misguided ecologists—earns J. L. Hudson's contempt.

"This year," he begins his introduction to the latest edition, "I would like to talk about the popular idea that 'non-native' species are somehow harmful, that 'aggressive exotics' can invade ecosystems and destroy 'native' species." Hudson claims there is no biological validity to these ideas whatsoever, and he gets rather testy on the point: "It is ironic to me to hear people of European ancestry accuse other organisms of being 'invasive exotics, displacing native species.' " (I would certainly think twice before uttering the word "weed" in Mr. Hudson's presence.)

"On the contrary," he continues, "the aid we have given species in their movement around the world has served to *increase both global and local diversity*. It is one of the few human activities which is beneficial to the non-human creation. It cannot be distinguished from the movement of species by wind or ocean currents, or the aid other species give to their fellows, such as the distribution of seeds by migrating birds."

For all of humanity's faults (and Hudson can be very hard on his own species), we are currently one of the most important vectors of "inter-species DNA transfer," a process that is vital to the progress of evolution. And even though Hudson tends to think in long, evolutionary terms, he speaks of our mission with the utmost urgency: "We have only a brief moment in history when fossil fuels will continue to allow us rapid worldwide travel. Let us use this time wisely." His text here is from William Burroughs: "Migrants of ape in gasoline crack of history."

J. L. Hudson's faith in the value of seeds strikes me as a noble, even touching thing. Who else would sell the seeds of common weeds (pardon me) such as mullein and stinging nettle next to the seeds of giant sequoias(!) or of the Mexican teosinte plant—the sacred wild maize that Indians have kept in their fields for thousands of years in order to reinvigorate their seed corn? Here in one catalog are the makings of an

English perennial border and the garden of a Zapotec Indian—more than fifty varieties of vegetables and ornamentals gathered by this Oaxaca people's last herbal healer. ("It is hard to describe the human effort that has gone into bringing these seeds to you.") The contents of Hudson's catalog form a vast, teeming democracy of seeds, the common and the rare, the useful and the useless, some of easy culture and others so stringent in their requirements that, he warns, it may take two years for them to germinate. But that's no reason to exclude them, not if you believe, as Hudson evidently does, that it is we who work for the seeds, and not the other way around.

Behind the pages of this seed catalog may stand a visionary. Hudson writes about the power of seeds with the reverence and awe that the Founding Fathers wrote about the power of words in a democracy. (Indeed, Thomas Jefferson, who once wrote that "the greatest service which can be rendered any country is to add a useful plant to its culture," would undoubtedly have admired and sought out J. L. Hudson.) Seeds have the power to preserve species, to enhance cultural as well as genetic diversity, to counter economic monopoly and to check the advance of conformity on all its many fronts. "Preservation through Dissemination" is Hudson's credo, a principle he raises (need I mention it?) high above profit: Hudson calls on his subscribers to save and exchange seeds with him and one another; he even gives the names and addresses of competing seed houses he esteems. You come away from Hudson's catalog convinced it was a serious oversight of the Founders to have framed the Bill of Rights without providing for freedom of seeds.

Each year the garden that takes shape in my winter imaginings is a little different, reflecting the tally of last season's successes and failures, as well as the wilder notions planted during my winter travels in the wider garden world—by my evenings spent among the catalogs. One year it was an English perennial border, planted in my mind by Amos Pettingill in January, and then replanted in my yard by me in May, though with decidedly less stunning results. The next year it was a border of antique roses inspired by Wayside's charming introductions and voluptuous photography.

Last January it was a European kitchen garden I plotted, tidy columns of leeks like I'd seen in pictures from France, frizzy heads of Lollo Rosso lettuce from Italy (via Cook's), a skyline of tomato cages festooned with golden Mandarin Cross tomatoes imported from Japan by Shepherd's, bunches of Mokum carrots, and heads of radicchio lined up like tight red baseballs along the paths. Some of this came to pass: I harvested enough Lollo Rosso to give bushels of it away in Manhattan; enough, I like to think, to have cut into Balducci's sales of the stuff at $5.98 a quarter pound. But the radicchio came in neither tight nor red; most of it proved too bitter to eat.

This year I'll continue to build on all three of these gardens—a big order from Wayside is scheduled to arrive the third week in April, and my vegetable seeds from Cook's are already sending up shoots on a sunny windowsill in the living room. But as you may have surmised by now, this winter my speculations have run to a garden of heirlooms. I have been made wild by the catalogs of Jan Blum and J. L. Hudson.

And this is what I'm thinking now: of my garden as a kind of blooming archive, a multicultural, transhistorical crossroad where Sibley squashes once cultivated by American Indians root in the same soil as the Madame Hardy rose, the white damask bred in 1832 by Monsieur Hardy, gardener at Malmaison. Welcome here, too, in this ordinary patch of Connecticut soil, will be the honeyed, green-fleshed Jenny Lind, beloved muskmelon of the nineteenth century, cast aside by commerce (just because, it's said, her rind wasn't tough enough to travel) only to be rescued by seed-savers like . . . well, like *me*! Incredible, don't you think, to have two such different women of the nineteenth century commemorated in one garden—the wife of Josephine's gardener, certainly one of history's faceless and forgotten, and then the "Swedish Nightingale" Jenny Lind, perhaps the most famous woman of her time, a soprano idolized on both sides of the Atlantic, the century's Madonna. It is said that her beauty and coloratura set millions to swooning. Come July when those melons start to swell I will be thinking of her—and of her bosom, I guess, for why *else* would someone name a melon after a sex symbol?

And here, too, rising like conning towers over my garden, will be hollyhocks I've grown from seeds that I traded my Sibley's for, and which my correspondent tells me were harvested from plants whose seed she obtained from a painter on Long Island who gathered *his* seeds from Monet's garden at Giverny—a complicated daisy-chain of seed dispersal reaching all the way from Giverny to Long Island to Manhattan to Cornwall, the seeds sped here not by birds or trade winds but by Toyotas and 747s—along, in fact, the "gasoline crack of history." And not a bad trade, either, a warty old Indian squash for a flower whose beauty caught the eye of Claude Monet.

What a place!—where French Impressionists sun themselves alongside American Indians, and celebrated melons sip the rain with pedigreed roses. And in the afternoon everybody will welcome a bit of shade provided by the pair of antique apple trees I've ordered from the catalog of the Southmeadow Fruit Garden, in Lakeside, Michigan: Ashmead's Kernel, a russet said to be the most popular apple in England during the eighteenth century; and Esopus Spitzenberg, a red, gray-speckled apple Thomas Jefferson swore by, and planted by the dozen at Monticello. And flitting and zipping overhead will be millions of delicate green lacewings and ladybugs, dainty mercenaries I've recruited (through the catalog of the National Gardening Research Center in Sunman, Indiana) to police the garden of aphids and other undesirables.

The orders are out—to Jan Blum in Boise, to the bug people in Sunman, to the apple man in Michigan—and already the mail has brought the first packets of seeds and a pouch of lacewing larvae which are cooling their heels in my refrigerator. What's coming in now, by parcel post and Fed Ex and UPS, are the packets of genetic information that will give my garden form. But not *only* genetic information, for there is culture as well as nature inscribed in these seeds: these Sibley squash seeds will divulge by their fruits the tastes and cultural practices of long-dead Indians; and the Jenny Lind seeds just now articulating their cotyledons on my windowsill will also articulate something of what the word *melon* conjured in the mind of a Walt Whitman or Chester Arthur, something very different, it seems, from what the word summons for us. Even the bug larvae in the fridge are encoded with lines of unintelligible but priceless data, the sum total of a few million years' evolutionary knowledge about the ins and outs of aphid-hunting.

And what does all this make me, the guy who's thumbing the catalogs, placing the telephone orders, writing the checks, and starting the seeds? A kind of librarian, maybe, whose job it is to gather and organize these volumes of information from distant points (distant in time as well as space), to shelve the roses of Second Empire France along this wall, the squash of American Indians on that. But librarian's not really right, not quite, because what I'm making here is no mere catalog, not just an archive of the old but something that, in its juxtapositions and conjoinings, has no precedent: something new. No Dewey decimals around here, and don't come looking for any library's hush or order. This place come summer'll be more like a buzzing marketplace, a teeming, polyglot free port city where all manner of diverse and sundry characters—immigrants drawn from near and far, past and present, East and West, upstairs and down—will meet and mingle and fuse in heretofore unimaginable combinations. I expect J. L. Hudson would feel right at home here, Amos Pettingill not at all; he'd probably judge the neighborhood a little too dangerous, so boisterous and motley he'd be nervously patting his wallet the whole time.

So where do I fit in? Well, nothing much happens around here without me, not at the start, anyway, because most of this garden's denizens—like most of the plants in cultivation—are as dependent on the gardener for their survival as the flowers are on the bee or the trees on the squirrel. Without us to collect and protect and replant their seeds, the beans disappear, the squash and corn and apples flat-out vanish. So that's another thing this garden is, a saving remnant, a kind of ark (call me Noah!) on which the genes of Jenny Lind are borne into the future.

Maybe Hudson's right and I should think of myself not as the master of this city of plants, but as its servant, too—vector of interspecies DNA transfer, conveyor of information across vast stretches of space and time, means to an undreamed-of evolutionary end. Buzzing from catalog to catalog, dispersing seeds through the U.S. Mail, bringing together far-flung genes in fresh combinations—I always thought I did these things just to please myself, but maybe it's not that simple. Maybe you should call me Bumblebee.

CHAPTER 12

The Garden Tour

I realize now that the day my father mowed his initials in our shaggy front lawn on Long Island, he was acting out what could be considered a primal American gardening scene. He was declaring to his neighbors: I reject the conventions you've given me for ordering this land (*my* land!); now watch me strike my own relationship to it. Writing in the grass with his Toro, my father was putting his stamp on the place and, almost as important, drawing a line against the inchoate wilderness that his lawn by then had become. He was making, in other words, a kind of garden, albeit not a pretty one. Most gardening, it seems to me, begins with such a gesture—on the one side against convention, and on the other, against the wilderness.

In America the first move seems to be the harder one. To go your own way with a patch of land—to "design" it according to your own taste or fancy—is to turn your back on the community, to flirt in this nation of lawns with antinomianism. Indeed, even to speak of aesthetics with respect to the land has historically been regarded here (except perhaps among the wealthy) as vaguely un-American, almost profane. For how could one man possibly improve on God's country? What hubris! If we are going to alter this land at all, we will do it together, as a congregation, for reasons not of beauty but utility, and preferably in a plain, Protestant style. So we rolled out across this infinitely various country a single American lawn, and that pretty much was the end of it.

To speak in one voice about the land—this is not to make gardens, not really. And indeed Americans have given that word, *garden,* a decidedly idiosyncratic meaning, as if to admit to a certain ambivalence about the idea. Soon after I started gardening, I noticed that a garden in literature and a garden in everyday speech were two entirely different things. A garden in books is always a place, somewhere you can enter and walk around (and set scenes in), yet in common parlance the meaning of the word has been oddly shrunken: it usually denotes a bed or patch that one can point to—here is my flower garden; over there is my vegetable garden. What everyone else in the world would call a garden we call simply, plainly, our "yard."

Gardens and even yards in America are not places for being in but for looking *at.* We admire our beds from the lawn, and arrange our unfenced front yards for the admiration of the street. What other possible purpose could "foundation planting" serve? Rather than create any habitable outdoor space (which is what the same planting out along the road would accomplish), it merely adorns the house, showing it off to advantage like the setting for a gemstone. Suburban America has been laid out to look best from the perspective not of its inhabitants, but of the motorist. As one landscape architect pointed out in the twenties, "the suburbs of American cities are said to be the most beautiful in the world —*to drive through.* Could there be a more eloquent qualification of praise than that final clause?"

I'm convinced that gardening—*real* gardening, not just putting in beds of flowers or tomatoes—begins with the removal of one's property from the motorist's gaze, with one's secession from the national lawn. This might mean throwing a hedge or fence around your yard, letting it go to meadow, or ripping out the grass and putting in something else entirely. But once you've done this, made the big break, prepare to feel very much on your own. At least that's how I felt when I finally fenced in my front lawn and tore out large swaths of it in back; exiled from the national greensward, I stood on suddenly unfamiliar ground, unsure of my next move. What do I do now?

I soon found out that there are remarkably few reliable guides or relevant models to follow. For all the hundreds of gardening books published in this country every year, only a handful take gardening beyond a concern with the plant, or the bed. As Edith Wharton once noted, American gardens seem to exist for the benefit of our plants and not, as it should be, vice versa. There are plenty of American books on how to grow perennials, and even a few on how to make pleasing arrangements of perennials in borders, but hardly any that discuss the principles of landscape design in terms relevant to a small plot of land. We are lost, it seems, without our lawns, and still feel uncomfortable talking about aesthetics—about the *look* of our gardens, and how that is achieved. From the text of nearly any English gardening book it is possible to glean a picture of the author's garden (which most times is beautiful and, for reasons of climate and economy, impossible to duplicate here). But even the closest reader of an American garden writer, such as Allen Lacy or Eleanor Perényi or Henry Mitchell, will come away with no idea what his or her garden actually looks like.

Why should we be so inhibited talking about the design of our gardens—about gardens as *places,* and not just collections of plants? Probably because morality has always been so thickly intertwined with gardening in this country. We are, after all, descended from a race of garden destroyers. The Puritans despised ornamental gardening and, in England, they wrecked many of the great Tudor gardens during their time in power. They regarded a designed landscape as an infringement on God's prerogative. To garden expressly to please the senses has always been considered vaguely decadent in this country, evidence, even, of antidemocratic tendencies. When wealthy Americans did make great gardens in the nineteenth century, they took pains to design them not as pleasure grounds but as "model farms." These Jeffersonian gardens purported to benefit the young republic by developing and introducing useful new cultivars; instead of ornamental plants, fruit trees—considered at the time the most "republican" of plants—predominated. "Magnificently ornamented grounds might have excited comment from Democratic Republicans," the historian John R. Stilgoe explains, "had not the trees borne fruit."

It's not easy to design a beautiful garden when you have moralists, Puritans, Democratic Republicans, and, today, even plants'-rights proponents looking over your shoulder. The absurdity of this situation was impressed on me not long ago, during a winter visit to the home of a well-known environmentalist in Brooklyn Heights. This man happens to be an avid gardener, and behind his Victorian brownstone he tends an enviable patch of sunny land. How had he chosen to arrange his yard? Perhaps feeling that the layout of an urban garden called for a certain formality, the environmentalist decided his garden would radiate out from a central point that would be visible from the living room. But rather than use a statue or a small pool, or perhaps even a sun dial, to establish this focus, he set down instead, dead center in the middle of his composition, for all the world to see—his compost pile.

Because I was unsure what his answer would be, I didn't dare ask if this was some sort of joke, a bit of irony he'd turned on himself. I doubt that it was. No, I suspect this was just another instance of moralism's triumph over aesthetics in the American garden.

I don't imagine ethical considerations will ever be entirely absent from the American garden, nor should they be. But I fail to see why we can't *also* attend to aesthetics, why our gardens can't tell a few other kinds of stories besides the usual morality tales. On the subway home from Brooklyn Heights, imagining what Capability Brown or Le Nôtre would have had to say about the environmentalist's garden, I resolved to try to look at my own garden from a slightly different angle that summer, to pay more attention to its overall design. How does one go about finding a path out of the American lawn that leads somewhere more interesting—and more pleasurable—than either the environmentalist's dour composition in compost, or my father's rather shrill "autograph" style? Whether or not the gardener is conscious of it, his garden cannot help but have a design, be it conventional or idiosyncratic, inherited or chosen. And that design is going to tell a story—about who you are, about your relationship to your neighbors on the one hand, and to the land on the other. It was time to give some thought to the kind of story a tour of my own garden would tell.

Seven years after we bought this place, the garden is finally starting to feel like a garden, and not just a collection of plants and beds and borders. In speaking about its design, I confess to a certain hesitancy. First, being American, I can't help but feel there's something a little pretentious (or is it decadent?) in talking about a parcel of nature in terms of enclosures and prospects, rectilinear and curvilinear form, the "function of water" and the desirability of "surprises" in the landscape. Second, this sort of talk about a place tends to conjure in the reader's mind a rather grand property, and mine is not—really, truly not—that. I stress the point because I am mindful of that annoying convention of so much garden writing: false modesty. From Vita Sackville-West's old columns in the London *Observer,* a reader who didn't know better might have gotten the impression that Sissinghurst was a decent-sized suburban spread rather than the castle that in fact it is. What passes in most English garden books for a "small country house" is in reality a mansion that goes for three or four million in my part of the country. Well, this place is a *real* "small country house," a still pretty funky twenties farm restored and gardened on a shoestring, without paid help (except for the kid who mows the lawn). This is not Sissinghurst.

That said, though, I've become convinced that the aesthetic issues garden designers talk about do, or can, apply to even our modest places. After spending some time this winter reading about and looking at pictures of the world's great gardens (this, along with paging through catalogs, being the gardener's proper winter work), I was surprised at just how much they had to teach me—about what I'd already done (often thoughtlessly), and what I still need to do (for I am still in the middle of making this garden, as probably I will always be). Not in the specifics—the linked, yew-walled rooms at Sissinghurst, the stonework in the Boboli, the prospects at Stourhead—but in the aims and senses of these places, in the spirit that informs them, we can find things that, writ small, belong in our garden. (English gardeners must have always known this, or why would middle-class cottagers have bothered to read Sackville-West's weekly newspaper dispatches from Sissinghurst?) Though Alexander Pope would probably choke to hear it, the design of this place reflects some of the advice he offered the Earl of Burlington in his famous epistle—that the gardener "consult the Genius of the Place" on all questions; that "He gains all points, who pleasingly confounds,/Surprises, varies, and conceals the Bounds"; that the gardener would do well to "Start even from Difficulty, strike from Chance." I used to read stuff like that and think it had about as much relevance to me, an American in possession of a few scruffy acres of land and a very slender wallet, as medieval manuals of chivalry or statecraft do. But approached in the right spirit (and not slavishly), our conversations with the classic garden writers and designers have something to teach us about discerning the genius of our own little places, and about how we might begin to devise some styles of gardening that will suit them, ourselves, and this country.*

The Genius of this Place: for me, that has meant chiefly two things, one historical (the place had been a farm), the other topographical. The lay of this land is too dramatic and, in places, too difficult to ignore. A garden will either make use of it or be defeated by it. Picture a piece of pie canted

* Besides Pope, I've profited especially from the writings (and designs) of William Kent, Francis Bacon, Horace Walpole, Joseph Addison, Capability Brown, Humphry Repton, and Vita Sackville-West, among the "classics." The most provocative contemporary guide to these figures is *The Poetics of Gardens,* by Charles W. Moore, William J. Mitchell, and William Turnbull, Jr. (Cambridge: MIT Press, 1988). Eleanor Perényi's *Green Thoughts* is also very good on the history of garden design (New York: Random House, 1981).

against a hillside; already you have some idea of the drainage problems I face. The broader end is at the base of the hill, and the land narrows to the northeast, rising more than a hundred feet from one end to the other. The hillside ascends rapidly, pausing only briefly now and then to dispense a small shelf of land. These planes, cleared by the farmer for cow pastures or outbuildings, seem almost to cling to the impatient hillside as if by their fingertips, giving the landscape an aspect of precariousness.

When we got here, only a single level—the one on which the house and a tiny front- and backyard sits—was being used; covered in lawn, it formed a small, flat suburban island in the midst of an otherwise roiling and ramshackle landscape. Everywhere else you looked, the farm seemed to be dissolving back into the hillside as the second-growth forest moved in, and the spread of brush and tumble of stone retaining walls began to erase the lines and stepped planes that the farmer had so painstakingly drawn. On almost every level stood some rusted piece of farm machinery and the weed-entwined ruins of an outbuilding: a chicken coop, a farm-hand's shack, a cow barn—eleven structures in all, most of them sinking back into the dense, green sea of foliage like scuttled ships.

It was all we could do at first to tend the small suburban island, to mow the lawn and put in a few herbs by the back door—the rest of the land was so turbulent and rank that even trying to think of the place as a whole defeated us. But we knew that, eventually, somehow, we would begin to reclaim it. We had not come here looking for suburbs; we'd both grown up in the suburbs, not liked them much, and it was the parts of this land that still recalled a working farm (the barns, the sequence of linked pastures, the old apple trees) that had attracted us to it.

The family from whom we bought this house, being the product of a different history, saw the place somewhat differently, to judge by the way they'd treated the land in the four years since buying it from the old farmer's estate. They'd come from New Milford, a town thirty miles to the south of Cornwall (and that much closer to metropolitan sprawl) where agriculture was eagerly being sacrificed to more profitable and up-to-date uses of the land. For them, growing up closer than we did to the barnyard's odors of manure and ruin, farming held little romance. (This was evidently true as well for the farmer's two sons, who live down the road from us in manicured suburban homes.) These days, around here, a farm said you lived in the past, by the sweat of your own brow, and in the shadow of probable failure. A modern, kempt suburban spread, on the other hand, proposed a shining destination—it pointed you directly toward the middle class. Ax to plow to lawn mower: this farm told pretty much the same tale of "progress" being told all over New England. On a place like this, you can measure the distance the owners have traveled from farming to middle-class respectability in acres of unproductive lawn. And by the time we got here, lawn had been spread over this place like a lid on its agricultural past.

Bit by bit we set about lifting that lid, and every time we did the farm's past would bubble up from the ground, like a return of the repressed. Part of the pleasure of making a new bed around here is not knowing what will turn up once you peel back a section of lawn and start to dig. We've unearthed the skeletons of dogs and deer, a plow, several decomposing tractor tires, a couple of rusty bowie knives, children's toys, patent medicine bottles, gallon jugs for applejack, farm tools, bullet casings and, oddly, lots and lots of teeth, presumably belonging to animals. The smaller, less disturbing artifacts line a shelf in the kitchen, where they help us to conjure the house's ghosts. But our archaeology has turned up items of a material value, too: the stash of composted cow manure I found near the barn, which did much to improve the soil of my vegetable beds; a small fieldstone patio by the back door, and a stepping-stone path by the front, both of which lay fossilized beneath a thick cloak of sod.

A place like this is a kind of palimpsest, and much of our gardening has been a process of laying bare the marks of earlier hands on this land. It is in these marks that we've discerned some of the genius of the place, and they've served as guides to our design of the garden. One of our first projects was to expand the fieldstone patio and expose the stone path that led from the front porch down to the road, bisecting the tiny front lawn. This path passes between two unruly clumps of hydrangea that had been mowed down by our immediate predecessor in his haste to subdue the landscape. The layout makes me think there may once have been a dooryard garden here, one of those small plots of flowers that the wife on a New England farm used to tend in the picket-fenced enclosure between front porch and road. As we nursed the hydrangeas back to health and tidied up the path, a picture of the old dooryard began to emerge from the featureless lawn, almost like a photographic print coming clear in a pool of developer.

The past—both the land's and the gardener's—exerts a certain pressure in any garden; indeed, many can be read almost as a form of historical or biographical commentary. It's possible to interpret the lawns we inherited as a renunciation of the farm's past, and my ripping out of those lawns as a repudiation of my own suburban past. This idea, that a garden comments on prior uses of the land, was well understood by Alexander Pope and his contemporaries as they set about inventing the picturesque style of landscape gardening in the eighteenth century—during that odd historical moment where the best minds of a generation were occupied with issues of garden design. The new taste for a more "natural" landscape—for unfenced gardens with long prospects, curved paths, and "serpentizing" waterways; for gardens, in other words, that strived to look nothing like gardens—was in part a reaction against agriculture's conquest of the English countryside, a process that was drawing to a close in Pope's time. It was the enclosure of the nation's fields in an orderly checkerboard of hedgerows (which William Gilpin, a popularizer of the new term *picturesque,* declared was "disgusting in a high degree") that inspired a nostalgia for earlier, more natural landscape forms. The more the countryside came to resemble a formal garden, the more gardens came to resemble countryside. Like the lid of lawn our predecessors put on this farm, the picturesque garden was a dissent against, or a swerving away from, agriculture; by organizing one's property for consumption rather than production, one could demonstrate his distance from the farm and its lowly social standing.*

The picturesque garden, though in many ways an artificial creation, was partly a work of restoration. The pastoral countryside that predated enclosure, a gently irregular landscape of pasture and coppice, was one of its models, part of the genius of the place Pope advised property owners to heed. (Lancelot "Capability" Brown earned his nickname for harping on basically the same theme—on the "capabilities" of a given parcel of land.) I've sometimes wondered what Pope would advise for this place—whether he would encourage me to remove it still further from its farm past, in the direction of picturesque lawns, or if perhaps he would look around at the local landscape and realize that the farm, at least today, here in New England, is special and rare enough to supply the gardener with an appealing and usable past. But Pope would probably declare in favor of the former approach; the aesthetic appreciation of agricultural land would have been as bizarre an idea to him and his contemporaries as the appreciation of forest land would have been to the Puritans, or to the man who used to farm this place.

Whatever his actual advice might have been, I'm inclined to lean on the spirit, rather than the letter, of Pope's first law of gardening, that spirit which can regard "farm" as the genius of this place, as the past worth recapturing. These days, the farms of New England are as endangered as open landscapes were in Pope's day, and it is possible, perhaps for the first time, to think in terms of what might be called an agrarian aesthetic. I'm hardly alone in finding beauty in a New England farm; most of us now would agree there is more beauty there than in the scruffy sprout-land and second-growth forest (not to mention the cookie-cutter subdivisions) that are quickly taking the farm's place around here. Gar-

* There was also a political subtext to the new aesthetic: among other things, the picturesque style represented a Whiggish reaction against the Royalist taste for foreign formal gardens; to plant a landscape garden was to strike a blow for English liberty. See the chapter on pruning in Perényi's *Green Thoughts.*

dens always seem to play against whatever constitutes ordinary landscape in their time, and that today is not farm but forest and, increasingly, suburb.

For me, here, making a garden with some thought to an agrarian aesthetic has meant, first off, a fair amount of restoration work. I've tried wherever possible to recover and emphasize the farmer's layout of the place, his paths and stonewalls and any salvageable structures. I've also worked to restore (generally by means of radical pruning and fertilization) what remains of his original plantings—the two hydrangeas, a dense thicket of lilac, forsythia, bridal wreath, and pepperbush out along the road, and a leggy old honeysuckle that's probably been loitering by the front door since the 1920s. A big chunk of our budget went to the tree man we hired to get the ancient apple trees back into shape; their intricately crabbed forms had gotten lost in a tangle of waterspouts and top growth. Today the apple trees once again bear fruit.

As we've worked gradually to extend the garden from the area immediately behind the house back toward the upper meadows, we've more or less followed the path that the farmer's cows took out to pasture every morning. That path, running from west to east, travels along a narrow plane that has become the garden's principal axis. Rising to your left as you walk back from the house is a fieldstone retaining wall that establishes the shelf of land on which the barn sits; it was along the base of this wall that we planted our first perennial border. Since the path journeys from the relative civilization of the back patio to areas of wilderness our gardening has yet to conquer, we planted the border to emphasize (really, to concede) this gradual diminishment of control: civilized and relatively delicate species such as aquilegia, veronica, and oriental poppies set out near the house gradually cede ground to tougher characters like rudbeckia, day lilies, mallow, evening primrose, and, finally, lythrum and artemisia, two plants quite happy to duke it out with the local weeds and brush unassisted.

To your right as you follow this path the land drops off eight or ten feet, where you can see laid out below the vegetable garden: five raised wooden beds set into the middle of a fenced rectangular lawn like a series of banquet tables in a big green room, each one with a dwarf apple tree holding court at its head. (On the far side of the fenced lawn lies a bog which we do our best to ignore.) The vegetable garden is accessible from the main path by a stone stair I dug into the steep bank, which, densely overgrown with goldenrod and vetch as it is, has the effect of rendering the two planes it divides all the more calm and hospitable. At least that's what I'll tell myself until I figure out what to do with it.

Every year we push the frontier of the garden a little farther out along this axis. At the beginning, the path just sort of petered out in a forbidding tangle of saplings and briars—not much of a destination for a garden walk. So a few years ago we hired someone to clear out the small trees and brush; now, looking down the path from the house, you can see clear through to our meadow and, rising above that in an even more intense shade of green, the well-kept hayfield of our neighbor, a field that was once part of this farm. Thus a chain saw and a machete gave us our first vista, a stairway of ascending pastures and hedgerows framed by a bent old white oak on one side and an ash on the other. I especially like the fact that we've managed to incorporate somebody else's field in our view, that we've appropriated his land and labor for our enjoyment. In this we've taken a page directly from the eighteenth-century designers, whose landscapes "leapt the garden fence," creating vistas that disregarded the bounds of a property in order to imply it was larger than it was. A chain saw has turned my five acres into ten, making this farm whole again, if not in deed then at least in prospect.

The picturesque designers—Kent and Repton and Capability Brown—made a point of hiding not only the boundaries but also any productive land from view, a prejudice we perpetuate in our suburban plots, the design of which in fact descends from the picturesque tradition. If a suburbanite has a vegetable or fruit garden, he will invariably relegate it to the backyard, leaving the front free to play its part in the picturesque park that the combined lawns of the neighborhood aspire to recall. Here, however, I've tended to keep my productive plots clearly in view, not out of any particular conviction, but because I happen to like the look of cultivated lands—the tight rows of vegetable seedlings lined up in their beds, the orderly parade of annuals in the cutting garden outside the kitchen window, the apple and peach trees set out in a grid between the house and road. What I like about such land is precisely what offended the classical garden designers: it betrays the human effort that went in its making, indeed invites us to take pleasure in the consciousness of skill applied. But if by incorporating cultivated areas in my design I am flouting romantic canons of taste, I am honoring the place's history and, in a way, my own.

I've dwelled up to now on the impress of the landscape's past on the look of a garden, but of course the gardener's own past exerts a pull at least as strong. In the *Sakuteiki,* an early manual of garden design written by a court noble in eleventh-century Japan (and paraphrased in *The Poetics of Gardens*), the author advises us in making a garden to consider "the lay of the land and water. Study the work of past masters and recall the places of beauty that you know. And then, on your chosen site, let memory speak." At least since the time when Nebuchadnezzar made the hanging gardens of Babylon to assuage his bride's homesickness for the hilly countryside of her childhood, gardens have been molded by our memories. Is there any garden that doesn't cast a backward glance, gather meaning to itself by allusion to the places in our past? It can be as slight as the scent of a lilac leading us back to a childhood sanctum, or as grandiose as the life-size model train set that Walt Disney re-created at Disneyland, but some private Eden shadows every garden. What could be more obvious about them, or as wonderful?

Perhaps it is this retrospective glance that explains why garden design is such a conservative art form. Three thousand years of garden making in the West have produced essentially three basic designs: the foursquare *hortus conclusus,* the open geometry pioneered by Le Nôtre in Renaissance France, and the picturesque or romantic garden invented in

Augustan England, which, incredibly, still stands as the last significant development—the state of the art. There have been countless variations on these themes, but little true innovation, an oddity of history that almost surely owes to the particular power of the past in gardens. Here is one place where radical novelty is the last thing we want, a place whose very essence seems to preclude modernism. Perhaps this is another cause of the American uneasiness about gardens: we have traditionally looked to the landscape for a way out of history—for a glimpse of the divine in the wilderness, or a fresh start out on the frontier. But gardens keep throwing us back on the past, on the ages of man.

I suspect memory speaks in my garden in even more ways than I am aware. But certainly it's the source of my fondness for productive land, and the reason I came fairly late to flower gardening. I don't have to tell you that the Eden in this particular head looks a lot like my grandfather's garden in Babylon, Long Island, or that the first bed I dug on this place was for vegetables. It must have been there, in *that* garden of Babylon, where fruit trees and vegetables shared the limelight with roses and rhododendrons, that I acquired some tropism inclining me toward an agrarian aesthetic. Coming into any garden, my eye always seems to pick out the ripe fruit first: the grapevines clutching their fistfuls of marbles, the cardinals perching in the tomato plants, the sweet swarm of peaches in the August orchard. How could Le Nôtre and Pope and Capability Brown ever have expelled such beautiful plants from the garden? ("What peasants!" they'd snicker to one another, nodding over at me and my grandfather.)

Almost from the beginning I had a vegetable patch large enough to ensure a surplus, and I did taste that pleasure my grandfather took in giving away produce by the bushel, that peasant pride. I confess, too, to occasional Jeffersonian fantasies of turning this place into a market garden ("Let's see. . . . If Balducci's can get $5.98 for a quarter pound of Lollo Rosso lettuce at retail, how many acres of it would I have to plant before I could quit the city and live off this land?")—and a *real* market garden, too, nothing like that heavily subsidized strawberry patch Jimmy Brancato and I planted on Long Island, which depended on my mother to buy up its surplus like an obliging Department of Agriculture.

Even now, a particularly fragrant strawberry will unexpectedly call that garden to mind, complete with a picture of Jimmy Brancato bent over his hoe. As everybody knows, it is not so much the eye that summons the gardens of childhood, but the nose. What memoir of childhood doesn't at some point turn on the scent of a sweet pea or a freshly cut lawn or boxwood hedge, to leap the fence of years? Here, I think, is the deepest spring of the past's power in the garden: gardening is one of the very few arts (cooking is another) to make use of the olfactory sense, to harness its uncanny knack for unleashing memory. Madeleines are everywhere in the garden (and surely Proust is its guardian spirit). For me the acrid chemical smell of Ortho Rose Dust still has the power to summon an August afternoon in my grandfather's garden. Not terribly romantic, but there it is.

Proust wrote somewhere that the reason beautiful places sometimes disappoint us in reality is that the imagination can only lay hold of that which is absent. It traffics not in the data of our senses, but in memories and dreams and desires. A garden will move us to the extent it engages the imagination as well as the senses. Among other things, a garden is a passage somewhere else—to the personal and shared past its scents evoke, to the distant places to which its forms allude. Gardens exist not only in the here and now, but in the there and then, too. The good ones seem to strike a pleasing or intriguing balance between here and there; unsatisfying ones err on one side or the other. A garden that's all there (a lawn, a zoo) is likely to feel imposed on the land, cold or abstract; a garden that's all here (like most "natural" or "wild" gardens) is apt to be slack or insipid, indistinct from the surrounding landscape.

Another way of saying the same thing is that gardens are simultaneously real places and representations. They bring together, in one place, nature and our ideas about nature. Like landscape paintings, gardens offer pictures of nature but, as the art critic Robert Harbison puts it, "unlike paintings, [gardens] let us forget there is anything beyond." There is, too, the difference in materials, between the plant—growing, changing and dying, insisting continually on its literalness—and the comparatively tractable tube of paint. Yet literal as it may be, that plant can, like paint, point beyond itself. A tree in a garden, as my maple taught me, is also a trope. But imagine: a trope that gives real shade.

At first my gardens here lacked this quality of doubleness, or resonance, and I realize now that is probably why they failed to make much of an impression—either on visitors or the land itself. In my mind, the vegetable beds may have sounded certain pleasing echoes, but to everyone else my gardens for the longest time seemed more or less invisible. That perennial border, with its long, gradual fade into the landscape, was so carefully contextualized as to almost vanish. (It didn't help that the local weeds promptly set about exploiting my accommodating style.) The surrounding landscape seemed to overpower all such subtleties; its predominance kept the gardens from ever rising above the here and now, from engaging the imagination. A garden should make you feel you've entered privileged space—a place not just set apart but *reverberant*—and it seems to me that, to achieve this, the gardener must put some kind of twist on the existing landscape, turn its prose into something nearer poetry. He has got to give it an inflection, and this was something I had not yet done.

Less as a result of planning than chance, I discovered that a certain degree of formality could begin to supply the missing element. By formality I don't mean pergolas and fountains, or boxwood hedges and parterres; *that* kind of formality would be a joke around here. What I really have in mind, I guess, is form: straight lines and rectangles, repetitions and occasional symmetries, a few simple passages of what Le Corbusier called man's language: geometry.

I think it was when I wisely abandoned the folly of my "natural" annual bed, following the triumph in it of the weeds, that I first understood the satisfaction of making a straight line in nature. I immediately liked the way a freshly cultivated row of plants stood out against the rolling land around it, the stillness of it in the face of so much upheaval. That rub, between the flat man-made line and the landscape's own bent toward curve and motion, seems to lend a certain energizing tension to a garden, to give it, quite literally, an edge. It begins to suggest that a place might have a story to tell.

Yet, benign as it might seem, I've discovered that the straight line in the garden is a controversial matter. After publishing an article recounting my travails in the annual bed, I received letters from both environmentalists and landscape designers criticizing me in the strongest terms for the formal layout I ultimately adopted. By "condoning existing esthetic conventions"—i.e., planting in rows, in a rectangle—I was acting "irresponsibly," a landscape designer from Massachusetts charged. My flower bed was contributing to the degradation of the environment, he claimed, because gardening according to "existing conventions" relied excessively on fertilizers, herbicides, and pesticides.

Now although it is true that the quest for a perfect lawn in certain places will require the heavy use of chemicals, this is less a function of aesthetics than geography—of the fact that lawns are often ill-suited to the American climate, and so must be coddled (with chemicals, etc.) if they are to look good. There is in fact no inherent reason why the most formal of gardens will necessarily be less environmentally responsible than a so-called natural one. On this point Eleanor Perényi is absolutely correct: "I object to the idea that only a replica of the wilderness can qualify as an ecologically sound environment." A "wild" garden is not intrinsically healthier, or even more preferable to nature, than a well-tended parterre. Whether a garden is ecologically sound or not depends solely on the gardener's methods; his aesthetics has nothing to do with it.

This romantic (and preposterous) notion—that nature somehow prefers one style of garden to another—was first advanced by the eighteenth-century landscape designers. "Nature abhors a straight line," William Kent declared, thereby dismissing every single previous garden (and, most pointedly, Versailles) in a scant five words. Like anybody with an ax to grind has done at least since Rousseau, Kent and his colleagues were invoking the unquestioned authority of "nature" to support their point of view, a tactic that always bears watching closely. It might sound good to say nature abhors a straight line, but is it really true? A plausible case could be made that nature *loves* straight lines, or else why would she be so enthusiastic about gravity? That falling apple didn't "serpentize" one bit en route to Newton's cranium. I'm inclined to think nature is actually indifferent on the subject of straight lines, that she doesn't care one way or the other whether I plant my asters in neat rows or free-form "drifts."

The picturesque designers made much of the "naturalness" of their landscapes, and that probably accounts for the strong impression they made on Americans like Andrew Jackson Downing, Frank Scott, and Frederick Law Olmsted, the three men largely responsible for the look of American landscape design. But for the English designers "nature" was not so much a divinely ordained model (what it would become in America) as it was almost a kind of conceit. Horace Walpole's wry accounts of the movement make this plain: "There is not a citizen who doesn't take more pains to torture his acre and a half into irregularities than he formerly would have employed to make it as regular as his cravat." To achieve their natural effects, the picturesque designers employed plenty of artifice; they just went to great lengths to keep it out of view. "Art should never be allowed to set a foot in the province of nature," Walpole advised, "otherwise than clandestinely and by night." Clearly this crowd takes us only partway to romanticism. For Wordsworth or Thoreau would never have commented, as one picturesque designer is said to have done upon observing for the first time the serpentine path of one stretch of the Thames: "Clever."

The perfect synecdoche for the picturesque garden—and its somewhat arch relationship to nature—is the "ha-ha," which also happens to be its enabling technology. The ha-ha, whose use in England was pioneered by William Kent, is essentially a fence laid in a ditch so that it will be invisible from afar and thus won't mar the distant prospect.

Without the ha-ha, the landscape designers could never have leapt the garden fence, or the cows almost certainly would have followed them.

As the name the eighteenth century gave to this ingenious device suggests, the picturesque designers never forgot that nature isn't necessarily natural, that occasionally it might have to be made rather than found. There is a certain playful irony in their fondness for nature that later romantics (and especially the Americans) would drop. No doubt because they stood in its vanguard, the early romantic designers understood that the appreciation of "wild" nature is in fact the most cultivated of tastes. As they well knew, it certainly did not spring full-blown from the contemplation of untouched land in England. The picturesque style, as the name implies, was inspired not by looking at land, but at landscape paintings, specifically those of Lorrain and Poussin. Both of whom, besides having been strongly influenced by classical writings about Arcadia, happened to be painting idealized scenes of the *Italian* countryside. Which happens to be one of the most thoroughly man-made landscapes in Europe.

And so it is that an American garden as seemingly "natural" as Central Park can be traced back from the highly sophisticated (and technologically intricate) nineteenth-century plan of Olmsted and Vaux to the eighteenth-century aesthetic theories of Kent and Pope to the seventeenth-century paintings of Claude and Poussin to the Arcadian poetry of ancient Rome. In such a dizzying regression of influences, it's hard to say exactly where nature itself fits in. But I think it is safe to say that the sort of righteous cant that attaches to the romantic style of garden design nowadays does not so much express nature's preferences and abhorrences as it does man's. Long live straight lines.

Since my first controversial foray into artifice in the annual bed, I have added several more geometrical passages to the garden, the most flagrant being the herb garden I recently planted behind the barn. Formality in garden design is usually thought to be aristocratic but, as Thomas Jefferson maintained, there could be a republican formalism, too, one based on the fair intentions of the grid, the rationality of the row and the rectangular field. Up to now, my geometrical gardens unconsciously honored these Jeffersonian notions (which fell from fashion under the onslaught of nineteenth-century romanticism), but this new garden seemed like the place to try something a bit more Old World.

When we rebuilt the barn a few years ago, the new structure occupied only the front half of the original fieldstone foundation. This left us with a perfectly square perimeter of stone behind it, roughly thirty by thirty. We filled it with topsoil to bring it to the level of the barn floor, creating a kind of stage that stands several feet above the adjacent land. (A good thing, too, because chaos reigns just below: briars, nettles, goldenrod, boulders.) In the middle of this platform, we had a mason lay a round brick patio about eighteen feet across: a circle in the square. And around the brickwork we planted ornamental herbs—lavender, catmint, lady's mantle, purple and Russian sage, two kinds of artemisia, calmintha, and mullein—and a handful of companionable perennials: foxglove, Siberian iris, hollyhocks, coreopsis Moonbeam, salvia, and ligularia.

Actually, we planted the herb garden twice. The first time, still gardening under the influence of a "natural" aesthetic, we assiduously avoided any symmetry in our planting, lest the garden look too artificial. We followed only the most elementary of design rules: keeping low plants in front, tall ones in back, and planting them all in large "drifts"—never less than three of a kind together in order to avoid the dinky effect that a lone, stranded plant can create. Here and there we'd repeat a particularly effective grouping (quartets of lavender, coreopsis Moonbeam, catmint, and lady's mantle worked particularly well), but basically the garden was a hodgepodge. Though the well-defined geometry of the stonework kept it from getting altogether lost, it nevertheless looked slack.

Last winter I bought a sheaf of graph paper, some colored pencils, and a compass and drew up a plan for replanting the bed. I decided that, with its square stone border and emphasis on herbs, this garden should probably be a good deal more formal. I'd read enough about the history of gardens by now to know my garden had something in common with those of the Middle Ages and early Renaissance, that it was a sort of *hortus conclusus* without the *conclusus.* These walled gardens, commonly planted in herbs prized for their mystical or medicinal properties (purely ornamental gardening didn't arise until the seventeenth century), offered a kind of refuge, a well-protected and scrupulously ordered place set apart from the dangerous and chaotic world that lay just over the wall. In medieval times, leaping the garden wall was the last thing anybody wanted to do.

These gardens, which frequently adjoined monasteries, were cerebral places—rather more hermeneutical than hedonistic. Every plant in them bore an allegorical significance and, much like the allegorical literature and painting of the time, the full meaning of these gardens was available only to the educated, to those who held the key. Rosemary stood for the fidelity of lovers (since it was thought to aid memory), sage for old age, bay *(Laurus nobilis)* for the laurel that crowns the poet, etc. In the same way that Boethius unlocked Dante or Chaucer, *Gerard's Herball* unlocked the *hortus conclusus.* By means of their rich allusiveness, these places transcended their cramped quarters, joined being here to the imagining of there—of distant, metaphysical realms. Though it has been largely forgotten since Le Nôtre, meditation is another way to leap the garden wall, one that is sometimes more appealing (not to mention more practicable) than conquering the whole countryside.

Since our herb garden adjoins my study and my wife's studio, and is a place we usually come to read or work through the knots in our work, the medieval herb garden seemed to offer a fitting model: a contemplative garden, walled off from the world, a good place to read and meditate. I drew up a plan emphasizing the elements that gave those classical gardens their appeal. If it was to offer refuge, my garden would need a wall, obviously, so I drew in a hedge of old roses around the perimeter, albas and bourbons that would eventually reach six feet and give a sense of enclosure without making the place seem claustrophobic. Closer in, immediately around the brickwork, I decided to limit the plantings to traditional herbs—lavender, lady's mantle, nepeta, sage, lamb's ear, and artemisia. These are all comparatively subtle plants—none bear especially showy flowers, but their foliage has character and many give strong, evocative scents. (The longer I garden, it seems, the less important a plant's flower is in my estimation of its merits—though I haven't yet attained the airy indifference of Russell Page, who dismissed flowers as so much "colored hay.") The colors of these herbs are muted, too, mostly grays and blues interspersed with a few of the cooler yellows and a shot of rich deep green here and there for contrast. Nothing too flashy or sumptuous, in other words, which is what you want in a garden intended as a setting for reading and meditation. Almost all of the plants are set out in a simple symmetrical pattern, nothing as intricate as an Elizabethan herb garden, but carefully balanced so that one side of the garden offers a mirror image of the other. Too much symmetry can get tedious, but there is so little of it in this landscape that I figured it would seem welcome here, a respite for the eye.

Last spring I translated the herb garden from paper to ground and, though it will be a few years before it fills in completely and the roses grow large enough to form a hedge, already this garden forms a distinct place, one that feels like no other on the property. The roundness of it, combined with the coolness and delicate textures of the herbs, lends a certain air of calm to the garden. But more than anything else, I think, it's the symmetry of this garden, the precise balance of its plants, that seems to still motion and suggest repose. To come on this place after a walk through the turbulent landscape all around it is to feel that you've made landfall at last, happened upon the most tranquil and orderly of islands.

Though they would never have approved of my little *hortus conclusus,* I suspect even Pope and Kent would have sympathized with my desire for a bit of order around here. They understood that a garden can be *too* irregular, and that aesthetic pleasure seems to depend on some measure of form, however slight or subtle. Kenneth Burke once defined form as a kind of rhythm in which expectations are aroused and then somehow paid off or fulfilled. He was talking about literary form, of course, but his definition applies equally well to the garden. Once begun, the garden path must take us somewhere, and then it had better bring us home again. The eye, after spotting a cool gray clump of lamb's ear in one corner of the garden, will then go off looking for its counterpart; if none is found, it will register a certain dissatisfaction—though we may never be conscious of the cause. Once the gardener has started something, in other words, he had better figure out a way to finish it. Symmetry is probably the most basic kind of form a garden can have—a sequence of visual expectations quickly and predictably paid off, as if the plants have been made to rhyme in couplets.

Such simple forms have an appeal all their own, but often, it seems, our pleasure is increased by the introduction of complexity and, specifically, of uncertainty—of some doubt as to whether the expectations that have been aroused will ever be fulfilled. To put form itself in jeopardy: this, too, seems to keep us interested, our minds alert. (Instead of the easy rhyming couplet, think of the slant rhyme, or of lines of blank verse sprung or enjambed almost to the point of no return.) Good gardens often seem to have this quality, of order under a certain amount of pressure, wilderness just barely contained. They make something of the fact that nature seems to resist our forms, turn this fact of fate to good account.

However imperfect the execution of it has been, I'm beginning to think that this may be the principal theme of my garden. A walk through it threads a path between those few areas of cultivation and so many wayward others—all those areas I have yet to bring under thumb. But now that I've seen how my orderly herb garden plays against the land around it, I'm no longer so sure I will do that, ever. Rather than tame the landscape entirely, it might be better to make something of its rankness—to take Pope's advice and "start even from Difficulty, Strike from chance." For whatever drama this garden possesses seems to turn on the tensions between these two kinds of space, between mulled and heedless ground—between the carefully constructed proofs of my labor (and of the old farmer's before mine) and the landscape's ceaseless drive to negate them. These margins may be what this place is about, its genius, and to bring to heel all that untamed territory—indeed, to finish my garden—might spoil the particular story it has to tell.

Capability Brown used to maintain that the most important kind of form a garden can have is an itinerary. It should unfold gradually, more like a narrative than a single picture such as at Versailles (all of which can be comprehended at a glance from the king's bedroom window). Instead of that one large impression—of royal power, which is the main story Versailles has to tell—there should be a succession of smaller ones: scenes of mystery, melancholy, romance, humor, and even sublime terror, all of these linked by the garden path. In order to "read" the picturesque landscape, we must venture out into it—go on a little journey. This notion of an itinerary strikes me as especially appropriate for the American garden, if only because the landscapes we seem to like best are ones we can do things in: places in which to explore, hit a ball, go on a ride, take a drive. Perhaps this explains why some of our most successful gardens (or at least the only ones the rest of the world cares to emulate) are golf courses and theme parks and parkway corridors.

It was when I figured out the itinerary of my garden that it finally began to jell, to feel like a garden. I had come to the conclusion that, as an organizing principle, the view out to the meadow wasn't enough by itself—that there had to be a good reason to walk that way, something to propel you beyond the perennial border and vegetable garden and deeper into the landscape's more dubious regions. I noticed that the picturesque designers always put some bit of artifice in the distance—a statue, a ruin, a little sign of art to beckon us down their paths. Kent used to call them "eye catchers." So down past the perennial border, where no one ever seemed to want to go, I built a small wooden arbor and covered it in clematis. And to make absolutely sure no one overlooked my eyecatcher, I erected above it, atop a six-foot pole, a white purple-martin house which is visible from just about every point on the property, like a steeple.

But I still needed a destination—some way to fulfill the expectations that a garden path leading through an arbor would arouse. Yes, you could pick your way from there out to the meadow up ahead, but I wanted to offer something a bit more surprising—something that would, in Pope's words, "pleasingly confound" the expectations set up by the view, which suggested a steady progression from garden to ordinary land. So just past the arbor, curving somewhat abruptly off to the left, I built a stepping-stone path and a handrail made out of saplings to beckon you up a small hill onto the plane where the herb garden sits. Up to this point the tour of my garden has ventured farther and farther out into less and less amenable territory—the perennials have given way to stinging nettles and elephantine burdocks, the wide grassy path has been replaced by a string of stones that skirts a marsh and even passes at one point beneath a precarious-looking jumble of boulders. Welcome relief, then, to come upon the orderly island of my herb garden, this still point amid the churn and slosh of green.

As the destination of this garden tour, the herb garden has the feel of a last-minute reprieve from the scruffy wilderness up ahead, an unexpected happy ending. All garden tours should end on some such note, it seems to me, for the transit of a garden should not be an adventure, not really, but a lark—more comic than heroic. A garden tour is, as the word suggests, a circuit, a circle that promises to bring us safely back to where we started. Along the way we're told a story—for the garden path is like the thread of a plot, or an argument.

And the story of this garden? Well, I suppose it's for others to interpret. But it does seem to have something to say about making a place in nature, about an optimistic man, an American, proposing a series of forms in the face of a heedless and difficult landscape. It's my story, of course, but it contains something of the old farmer's, too. More than anything else, this garden seems to tell the story of its own making. Which is why, I guess, it doesn't hide its labor (there, off to your right as you emerge from the arbor, tucked under the boughs of an ash, sits my steaming compost pile) and doesn't look to be anywhere near finished—and is probably also why it should end in such an emphatically gardenish place. To wind up in marsh or woodland would have been too melancholy, at least for this place and this particular gardener; it would have implied that a garden, that hoped-for reconciliation, is finally impossible.

A story about people and the land that ends happily—*that* form, which is the form of the garden tour, should never be in any serious doubt, though pretending it is can certainly keep things interesting. In my garden, of course, the pretending has always threatened to get a little out of hand. I suspect the "wild" areas in this place would have struck Pope and his friends as uncomfortably literal. ("Must be careful using terms like *natural* and *wild* around these Americans—they're liable to take you at your word.") Because that wasn't just a sand trap back there, some tame trope of trouble, but a real swamp with actual bugs and snakes.

I sometimes wonder what my grandfather would have thought if he could take a tour of my garden today. I suspect he would have been troubled by a lot of it, too. In fact, the only part I can be sure would pass muster with him is the vegetable garden. Especially after my early efforts on Long Island, the boxed, weedless geometry of my kitchen garden would definitely have gladdened him. But knowing my grandfather, his attention wouldn't rest among those scrupulous rows for long. He'd look over at the bog, right there on the other side of the fence, and want to know my plans for it—was I going to drain it and plant bluegrass, or dig a pond? I don't know how I would have explained to him that I planned to leave it just the way it is. That I'd grown to like the contrast the bog and garden make, the play between the disciplined heads of cabbage lined up two-by-two on this side, as if for some class trip, and the ungovernable mob of Joe-Pye weed and detonated cattails milling over there. About now, I expect he'd make the cash offer: to buy me a pond. Or a lawn—*anything* but this swamp.

I'd thank him for his generosity, and then maybe try another tack.

Something about the importance of leaving wetlands undisturbed, because of the complexity and fragility of their ecology. I would want to tell him that I thought that, nowadays, it might be a good thing for gardeners to go a little easier on the local wilderness, when they are lucky enough to have some of it left. And that maybe in doing so we would find a way to make our gardens a little less derivative—more American, and perhaps more contemporary, too. Not by giving them over to the forest or formlessness—that doesn't accomplish anything—but simply by taking some care to honor the landscape's past, and whatever remains wild in it.

The gardener knows how tenuous his control of nature is, especially here in North America, where the land can seem so ungovernable. So why then does he go to such lengths to hide this fact, to clothe such recalcitrant land in so much lawn? Maybe it's time we began to acknowledge, perhaps even evoke, that tenuousness in the design of our gardens. By leaving some parts wild, and by making a virtue of their juxtapositions with more formal areas, we can introduce into our gardens a measure of doubt about our control of nature, and that might be a good thing to do. But it's not more romance about wilderness we want, that's not what I'm saying—rather, it's irony about gardens, about these places we make in nature. In the same way that the picturesque designers were always careful to include some reminder of our mortality in their gardens—a ruin, sometimes even a dead tree—the act of leaving parts of the garden untended, and calling attention to its margins, seems to undermine any pretense to perfect power or wisdom on the part of the gardener. The margins of our gardens can be tropes too, but figures of irony rather than transcendence—antidotes, in fact, to our hubris. It may be in the margins of our gardens that we can discover fresh ways to bring our aesthetics and our ethics about the land into some meaningful alignment.

I know, I know—my grandfather'd think this was the biggest, fanciest bunch of excuses for lazy gardening he'd ever heard. But I'd want at least to try to make him see it, the tension I've come to love about this place, the edge it gets from the partialness of my control. Look at how all of the forms I've introduced here—the dead-straight lines and walls and arbors, the symmetries and repetitions and paths—seem to profit from the contrast with the rougher land all around them, acquiring a savor they might not otherwise have had—a poignance, almost, which is something that the same forms don't usually have in Old World gardens. Maybe this sort of irony is what the American landscape, and our time, has to give the garden, what we have to add to its story.

But it's not only the garden that gains from the rub against rough land—what's even more surprising is how the rough land seems to profit from the contrast as well. That cataract of boulders and crazy brush that lies directly beneath the formal herb garden has been rendered much more interesting by the juxtaposition it now finds itself party to; it could almost be the treacherous waters out beyond the safe harbor of this garden, the world outside the *hortus conclusus.* Even the formless, quaking bog has acquired fresh distinction by its proximity to the orderly schoolroom of my vegetable garden. Perhaps even wilderness depends upon a frame, upon the foil of human artifice.

The simplest, best proof of this idea I know can be constructed with a lawn mower and a patch of overgrown grass, the wilder the better. To look at a freshly mowed meadow path, the way it draws such a crisp, syntactical, human line through the soft and billowy, heedless grass, is, I think, to understand the gift of the garden to the wilderness, and its dazzling reciprocation.

This, for me at least, is a fairly recent discovery. After we put in the hedge along the road, we decided to let our big, south-facing lawn go to meadow. We had considered doing something more ambitious with it, but over time it became clear that our gardens should probably move back from the house, up the hillside, rather than down toward the road. In that direction lay greater privacy and, anyway, gardens that unfold upland seem to have more interest than ones that slope down. For this particular place, the lowest and wettest corner of the property, a meadow seemed like the best approach.

In the beginning, though, the meadow was a disappointment. The taller the grasses grew, the more slovenly the corner looked—more like a vacant lot, or my father's derelict lawn on Long Island, than the country meadow I'd had in mind. Whatever it is that makes a common weed patch into a meadow was lacking here. Yet I was determined not to resume mowing. Having so noisily forsaken the American lawn, I wasn't about to come begging its forgiveness, to ask to be taken back in. There could be no turning back, at least not for this gardener. So it was a lucky day for me when I discovered that I could put the lawn mower blade on the highest setting and cut a path through the tall grass that, at a stroke, transformed that sorry patch of grass and weeds into something altogether different—into a meadow.

That path, in my eyes anyway, is a thing of incomparable beauty, especially right after it's been mowed. I don't know exactly what it is, but that sharp, clean edge changes everything; it makes a place where there wasn't one before. Where before your eye sort of skidded restlessly across the tops of the overgrown grasses, in search of some object on which to alight, now it has an enticing way in, a clear and legible course through the green confusion that it cannot help but follow. The path beckons, making the whole area suddenly inviting. (Even my cat, whom the tall grasses never bothered, now makes a point of keeping to the path.) New possibilities have opened up: there's now the prospect of a little journey.

In a path is the beginning of narrative, that sure and welcoming sign of human presence. But it's not, at least here, a vaunting or brutish sign: this is a footpath mowed in my front yard, after all, not a highway through a forest, or my father's angry grass graffiti. The story this path has to tell is not about man battling nature—this is no tale of conquest or submission. A path through the grass is an entirely different thing from a lawn. True, I have to mow it again every week, and yes, the grasses seem intent on erasing my cherished line. But still this is no pitched weekly battle, Mower versus Grass. It's more like the argument of old friends, or husbands and wives after long years, a quarrel renewed week after week with no end in sight, and no end sought.

Am I making too much of my meadow path? Perhaps. But the longer I mow it, the more I can see my whole garden in that simple footpath, in the way that it seems to inflect the land, to give it something of ourselves without diminishing it. On days when I've mowed, the path seems as lucid and convincing as geometrical proof, like some fine, Apollonian line drawn against all that's inchoate or polymorphous in the world, a stay against entropy, a proud declaration of identity in the face of so much grassy indifference. Here is *my* autograph, then, scratched on this green page with my roaring Toro, this very prow of culture! This stylus of Western will! Yes, the grasses have their days, too, coming usually after a drenching rain, when the ink on the signature blurs and the fresh green growth dissolves my path's keen, whetted edge. No more than nature herself, these grasses haven't the least regard for any human form or identity, for even the greenest of our thoughts. But so what?! I can remow my path every week, just as I reweed my beds and borders, renewing these human forms, this entire garden, again and again, in the face of all indifference. Just because nature is bound to have the last word—will be forcing fresh shoots out of the ground long after I've ceased to weed or mow—is no reason to close the conversation now. Nature does tend toward entropy and dissolution, yes, yes, but I can't help thinking she contains some countervailing tendency, too, some bent toward forms of ever-increasing complexity. Toward us and our creations, I mean. Toward me and this mower and the otherwise unexplainable beauty of a path in a garden.